Textile Fiber Microscopy

Textile Fiber Microscopy

A Practical Approach

Ivana Markova
San Francisco State University
USA

This edition first published 2019
© 2019 John Wiley & Sons Ltd

All rights reserved. No part of this publication may be reproduced, stored in a retrieval system, or transmitted, in any form or by any means, electronic, mechanical, photocopying, recording or otherwise, except as permitted by law. Advice on how to obtain permission to reuse material from this title is available at http://www.wiley.com/go/permissions.

The right of Ivana Markova to be identified as the author of this work has been asserted in accordance with law.

Registered Offices
John Wiley & Sons, Inc., 111 River Street, Hoboken, NJ 07030, USA
John Wiley & Sons Ltd, The Atrium, Southern Gate, Chichester, West Sussex, PO19 8SQ, UK

Editorial Office
111 River Street, Hoboken, NJ 07030, USA

For details of our global editorial offices, customer services, and more information about Wiley products visit us at www.wiley.com.

Wiley also publishes its books in a variety of electronic formats and by print-on-demand. Some content that appears in standard print versions of this book may not be available in other formats.

Limit of Liability/Disclaimer of Warranty
In view of ongoing research, equipment modifications, changes in governmental regulations, and the constant flow of information relating to the use of experimental reagents, equipment, and devices, the reader is urged to review and evaluate the information provided in the package insert or instructions for each chemical, piece of equipment, reagent, or device for, among other things, any changes in the instructions or indication of usage and for added warnings and precautions. While the publisher and authors have used their best efforts in preparing this work, they make no representations or warranties with respect to the accuracy or completeness of the contents of this work and specifically disclaim all warranties, including without limitation any implied warranties of merchantability or fitness for a particular purpose. No warranty may be created or extended by sales representatives, written sales materials or promotional statements for this work. The fact that an organization, website, or product is referred to in this work as a citation and/or potential source of further information does not mean that the publisher and authors endorse the information or services the organization, website, or product may provide or recommendations it may make. This work is sold with the understanding that the publisher is not engaged in rendering professional services. The advice and strategies contained herein may not be suitable for your situation. You should consult with a specialist where appropriate. Further, readers should be aware that websites listed in this work may have changed or disappeared between when this work was written and when it is read. Neither the publisher nor authors shall be liable for any loss of profit or any other commercial damages, including but not limited to special, incidental, consequential, or other damages.

Library of Congress Cataloging-in-Publication Data applied for

Paperback ISBN: 9781119320050

Cover Design: Wiley
Cover Image: courtesy of Ivana Markova

Set in 10/12pt WarnockPro by SPi Global, Chennai, India

10 9 8 7 6 5 4 3 2 1

Contents

Preface

Textile Fiber Microscopy offers an important and comprehensive guide to the study of textile fibers. It contains a unique approach that prioritizes a review of fibers' microstructure and macrostructure. This book is written for students and professionals in textile science, and forensic science fields. *Textile Fiber Microscopy* presents an important review of textile fibers (plant, animal, regenerated, and synthetic) from a unique perspective that explores fibers' properties (such as comfort, wicking, and absorbency) through the understanding of fiber morphology. The text is accompanied with a number of micrographs, both black-and-white and colored. The micrographs are to enhance the understanding of fiber structure and also to encourage students to use microscopes. The field of microscopy is briefly described in the introduction of this book, and the preferred microscopy tools are recommended for a variety of fibers throughout the text. This book provides a comparative textile fiber review that facilitates deeper understanding of the material. Micrographs and diagrams have been carefully selected to illustrate concepts necessary in understanding textile fibers, their properties, and their appropriate end use.

Contemporary issues of environmentally friendly practices in fiber production are also incorporated in this text. The text includes a review of environmentally friendly fibers and contains information on some current fiber science by putting the focus on fibers that have been mechanically or chemically recycled, for use in textile production. The author also offers an exploration of issues of textile waste and the lack of textile recycling.

However, fiber science would not be complete without the mention of historic fibers and how microscopy makes identification of fiber artifacts possible. This book explores textile artifacts from Ancient Egypt, Ancient Greece, and the Carpathian region in Eastern Europe. This text is also appropriate for forensic scientists as it describes fiber shapes in detail, including fiber length and diameter measurements. The comparative aspect of this text will guide novice forensic scientists in unknown fiber identification.

December 18, 2018

Ivana Markova
San Francisco State University
San Francisco, California

Acknowledgments

I want to thank my family and friends for continuous encouragement and support.

My gratitude goes to Victoria Yao-Hua Lo for adding color design to micrographs and making them come to life, and to Mana Markova for her assistance with sketching graphics. I am grateful to Judy Elson, a textile expert, for her assistance with microscopy. All of your patience is greatly appreciated.

Introduction

Imagine a flannel robe against your skin, so soft that you can hardly feel its touch. Tightly wrapped, you are at ease, surrounded by comfort and warmth. The softness your robe provides is born of the cotton fibers making up the flannel fabric. Cotton fibers are flexible, convoluted strands which, when woven together, create a material perfectly suited for wrapping around the body. While you may be able to feel the effects of the cotton fibers in your flannel robe, they are impossible to see with the naked eye, and their coiled shape is visible only under a microscope.

Microscopy is an irreplaceable tool in the identification of textile fibers. With a powerful lens, it is possible to observe the characteristics of individual textiles. While the microscope has been around for some time, students still find the process of seeing the textile world up close fascinating. Dating back to the seventeenth century, the microscope has evolved to become an important tool in scientific observation. Cornelis Drebbel, Zacharias Janssen, Galileo Galilei, and Robert Hooke are some of the scientists credited with the invention and development of microscopes. Robert Hooke's book, *Micrographia*, published in 1665, depicted his microscopic observations and was one of the best sellers of that time. However, the adaptation of microscopy was greatly impacted by Antonie van Leeuwenhoek (1632–1723), a Dutch fabric merchant. Referred to as "the Father of Microbiology," he was neither a biologist nor the inventor of the microscope, though he is responsible for some of the greatest improvements to the tool. Prior to Leeuwenhoek's microscopes, microscopic images were distorted and hardly captured the details of what was observed. With the release of his improved microscope, biologists and scientists of the time hardly believed what could be seen. He handmade each microscope and inspired the creation of some of the first hand-held microscopy tools (see Figure 1). Most notably, Leeuwenhoek is known for keeping a detailed record of his findings. He drew sketches of tiny organisms, which he titled *animalcules* that we call microorganisms today. Leeuwenhoek and his microscope were the first to explore the microscopic aspects of the world we live in, studying everything from the size of bacteria to the blood flow in small vessels [1]. Antonie van

Figure 1 Antonie van Leeuwenhoek started his career as fabric merchant and later inspired the creation of hand-held microscopy. *Source:* Reproduced with permission of National Academy of Sciences.

Leeuwenhoek's work was amazing, but as with any new scientific observation, true biologists were skeptical.

When Leeuwenhoek was only 16, his mother arranged for him to begin an apprenticeship with a Scottish cloth merchant in Amsterdam. This became the first place he used a simple magnifying glass. While it could only magnify 3×, he was absolutely fascinated by the viewing and identification of fabrics and fibers. The fabrics were yarn-type and woven, and Leeuwenhoek learned that a close examination of a fiber under a magnifying glass could reveal a great deal about the fabric's properties.

A cloth merchant's primary responsibility was to closely check fabrics and determine their quality and value. In the seventeenth century, there were no manufactured or synthetic fibers. The only fabrics on the market were made of natural cellulosic or protein fibers. The cellulosics seen were primarily linen, cotton, hemp, nettle, and jute, and the proteins were wool and silk. To tell cotton from linen, or high-quality wool from low-quality wool, a cloth merchant needed a closer look. Antonie van Leeuwenhoek's curiosity grew out of this textile observation process. He would inspect fabrics for damage by mold or other infestations, or note the quality of dying. If he finds that a dye had not fully penetrated through the yarn or fibers, then the quality of the fabric would be deemed worthless. Becoming a cloth merchant required a deep understanding of textile fabrics, typically acquired over time through an apprenticeship. Working with textiles was a challenging job, and required proper training, even in the seventeenth century. The experience Antonie van Leeuwenhoek acquired in his lifetime allowed him to construct lenses and microscopes that permanently changed microscopy. While he never revealed his methods of creation, one is sure to remember that he was not only a great tradesman but also an amazing scientist and craftsman.

The microscope, as we know it today, has greatly advanced because of Leeuwenhoek, with amazing improvements in the nineteenth century, including the development and adaptation of the lens. An important contributor to lens development is Carl Zeiss, a German mechanic who partnered with scientists Ernst Abbe, a physicist, and Otto Schott, a glass chemist to create a better resolution technique. The heightened resolution improved the quality of microscopes, inspiring extensive improvements during the past 200 years.

1 Types of Microscopes Used in Science

Today, the microscope is commonplace, a simple instrument present in every laboratory. However, microscopes have come a long way, and their viewing and functioning properties have become quite complex. A variety of microscopes are used for specific purposes in scientific laboratories. Most of these use photons to form clear images and are called light microscopes. Electron microscopes, specifically the scanning electron microscope (SEM), are used in large-scale, full-service laboratories. These microscopes have a massive range of magnification allowing scientists to analyze fibers in a way that light microscopes cannot. SEMs have a very high resolving power and the ability to perform elemental analyses when equipped with an energy- or wavelength-dispersive X-ray spectrometer.

Microscopes can be differentiated by comparing the images they generate. The physical principle utilized by a microscope is equally as important, as it will usually determine why fiber images differ when viewed using different microscopes. Different microscopes visualize different physical characteristics of the sample. Resolution and magnification, which will be explained later in this section, are to be taken into consideration. The most common magnifications used by students to enlarge a fiber image are 4×, 10×, 40×, and 100×.

1.1 Stereomicroscope

The stereomicroscope is one of the simplest and easiest types of microscopes to use. It works by bouncing the light off the surface of the specimen rather than transmitting it through a slide. They are primarily suitable for observations not requiring high magnification. Its low magnification power (ranging from 2.5× to 100×) is due to its design. This microscope's illuminators can provide transmitted, fluorescence, brightfield, and darkfield reflected imagery, which allows the viewing of microscopic features that may otherwise be invisible.

With the stereomicroscope, there is a large gap between the specimen and the objective lens, which provides an upright, unreversed image. This space allows for better specimen manipulation and for a basic microscopic analysis to serve as the perfect preparation for a future, more detailed, microscopic examination and analysis. One important advantage of this scope is that the specimen does not require any special or lengthy preparation prior to observation. The specimen is simply placed under the lens and observed as needed.

The stereomicroscope is well suited for use in the preliminary identification of fibers, yarn, and weave structure when observing dated textile pieces for conservation practice. In general, textile fibers must be extracted from a yarn for proper observation and identification, but in viewing and identifying old textiles, such as tapestries or fabrics preserved for many years, removing fibers would damage the piece. With this microscope, the entire untouched, unraveled piece may be viewed without damage. In addition, this piece of equipment can be attached to a separate boom stand, allowing movement over a large object for examination. If a conservationist wants to examine the fibers of a new museum tapestry piece, a video camera may be attached to this microscope for proper record-keeping. Later chapters will include the conservation of textiles.

1.2 Compound Microscope

The compound microscope, also known as the optical or light microscope, uses light and a series of lenses to magnify particularly small specimens. Compound microscopes were invented in the seventeenth century and vary greatly in simplicity and design. These microscopes can be very complex and are a considerable improvement from the aforementioned stereomicroscope. While stereomicroscope can only magnify up to around 100×, compound microscopes rise in resolution and magnification up to 1300×. Today, the use of reflected light in microscopy outweighs the use of transmitted light. Regarding fiber examination, light microscopes are suitable for the analysis of fiber anatomy in hair fibers, such as the different types of medulla.

1.3 Polarizing Light Microscope

The polarizing light microscope is undeniably an advanced and versatile piece of equipment. It is normally equipped with a round, rotating stage, a slot for the insertion of compensators, and a nosepiece. It stands out from other microscopes due to its preciseness in both qualitative and quantitative fiber analyses. It embodies the functionality of normal light (brightfield) microscopes while allowing the researcher to view fiber characteristics transmitted through polarized light, as opposed to reflected light. In polarized microscopes, two filters are used as an illumination technique, also known as crossed polars. One filter, an analyzer, is placed above the stage, and the second filter, a polarizer, is placed below the stage. In this polarizing technique, the filters are crossed, and an effect known as extension or black out occurs. The fibers appear bright against a black background. Polarized microscopy utilizes contrast-enhancing technique to create a better image.

1.4 Electron Microscope

Electron microscopes are more sophisticated microscopes using electrons to form an image of the sample. SEMs are widely used in the textile laboratories. The SEM scans the surface of a sample with a focused electron beam to generate an image, converting the emitted electrons into a photographic image for display. This allows a high resolution and greater depth of focus. The SEM looks only at the surface of the specimen, which makes sample preparation simple. Instead of mounting a sample on a glass microscope slide, the specimens are placed on a strip of conductive tape that is attached to an aluminum mounting stub. SEM and the environmental SEM are primarily used in the identification of archeological textiles where detailed fiber morphological distinctions are required (Dennis Kunkel, personal communication, May 2016. Microscopy expert).

2 Magnification

In the study of textiles and fibers, magnification is extremely important. For student use, 10–40× magnification is typically sufficient for proper identification of fiber characteristics. The smallest magnification on a compound microscope is 4×, which allows students to pull their fibers into a focused view, but is not sufficient for identifying the actual fiber morphology. Once focused under 4× magnification, students can easily move the objective to a higher magnification to view the fiber characteristics. Even though most light microscopes have 100× magnifications, any magnification higher than 40× will be too close for students to view fiber characteristics clearly.

3 Resolution

Resolution, like magnification, is extremely important in microscopy. Resolution is a basic function of any microscope and represents the focusing power of a lens. A lens that can magnify an image without increasing the resolution provides only a blurry image and no specimen details. In reality, the resolution of a lens may be more important in microscopic analysis than magnification. In a good microscope, the resolution will increase as the magnification increases, allowing for clarity of observation and the viewing of detailed sample characteristics.

4 Use of the Microscope

Examine the different parts of a light microscope (see Figure 2). As the examination of fibers will utilize microscopy often, the following are some basic instructions provided for those students with no prior experience using the instruments:

1) When lifting or moving the microscope, pick it up by the limb or arm.
2) Never work in direct sunlight.
3) Use a firm steady table. The most comfortable seat for working with a microscope is a stool that can be adjusted to a comfortable height for viewing.

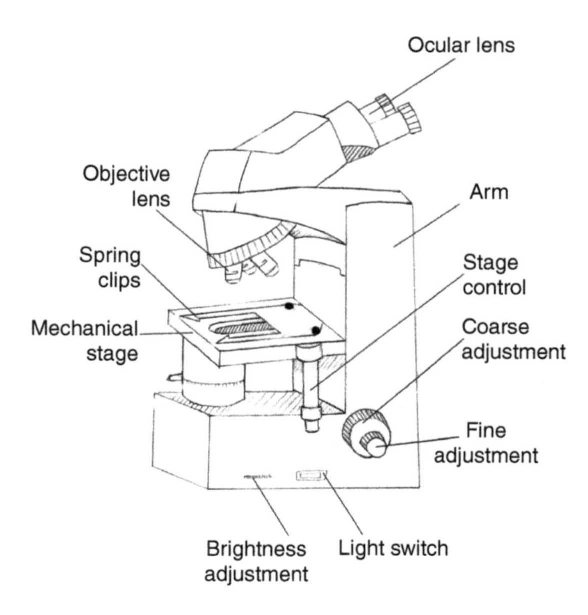

Figure 2 Microscope and its parts.

To prepare a slide:

1) Make sure you have a clean slide and slide cover.
2) Place a drop of water on the slide and add several fibers. Make sure you do not have too many fibers, as this can result in a crowded slide and identification of fibers becomes impossible.
3) Place the slide cover on top of water and fibers. Always be gentle with the slide covers as they are very thin and break easily.

To view the prepared slide:

1) Raise the microscope as high as possible.
2) Place the slide on the stage, with the fiber(s) centered over the opening for the light. Fasten the slide in place with the spring clips on the stage.
3) Lower the microscope until the objective is just a few centimeters above the slide. Do not allow the objective to touch the slide.
4) Look into the microscope. Turn on the illuminator or adjust the mirror, allowing the maximum amount of light into the microscope.
5) Start raising the microscope with the coarse adjustment knob. As soon as the fibers come into view, switch to using the fine adjustment knob. With the fine adjustment knob, pull the fibers clearly into view. Always focus the microscope by moving the objective up, never down, as lowering the objective may cause it to touch and break the slide or damage the microscope lens.
6) If you wear glasses, remove them for viewing. You will be able to adjust the focus to your eyesight.
7) Always look through the microscope with both eyes open. If you find this difficult, begin by placing your hand over one eye while observing with the other. Keeping one eye closed will cause fatigue over time.

4.1 Microtome

When attempting to view a cross-section of a fiber, the fiber must be cut into thin sections allowing light to pass through them. To cut fiber sections, you will use a fiber microtome, a tool specifically used for sampling thin cross-sections of all types of fiber. The microtome allows better microscopic observation of the fiber tissue structure. Microtomes, used specifically in microscopy, are similar to any instrument used for sectioning thin materials. Microtomes use blades that are typically made of steel glass or diamond. Blades of steel, for light microscopy, and of glass, for light and electron microscopy, are suitable to prepare animal, plant, or synthetic tissue for viewing. Diamond blades are primarily used to slice hard materials such as teeth, bones, or plant matter, not fibers. Microtome sections can be sliced as thin as 50 and 100 μm.

4.2 Measuring Fibers Using the Metric System

Instruments such as the microscope help us to see individual characteristics on fiber materials that cannot be seen with the naked eye. To measure fibers, scientists normally use the metric system. The metric system uses meters as the standard measurement of length. One meter is equal to 100 cm, and a centimeter is about the length of a fingernail. A normal cotton fiber is 1.5 cm long. Centimeters are further broken down into millimeters; 1 cm is equal to 10 mm. One millimeter is a very small measurement, and although we can still plainly see a single millimeter, it exists as the beginning of the microscopic scale. Scientists measure the length of fibers in centimeters and millimeters and the diameter in microns. The diameter of each fiber determines the fiber fineness. For reference, the diameter of human hair is about 1 mm. Most of textile fibers are smaller than a millimeter.

4.3 Sampling

When collecting fiber samples and preparing them for microscopic examination, one must remember to obtain a representative sample of the fibers to be viewed. Obtaining a small sample size limits the observation and may not yield accurate results. A sample must contain "notoriously variable materials," especially in examining natural fiber contents [2, p. 5]. It is important to examine multiple fiber samples to get the widest identification of its contents.

It is suggested that when dealing with blended fabrics, a preliminary sampling should be conducted to gain a truly representative sample for viewing [2]. For example, the observer should pull fibers out of both the warp and weft direction of the fabric without any selvages, treating them as separate samples. In a similar manner, yarns of varying colors should be examined by taking a separate sample of each color.

The most common specimen mountant used in textile laboratory is water. Mountants can be either temporary or permanent, water being the most temporary due to evaporation. A microscopic slide with water mountant cannot be stored. Water not only holds the fiber sample and slide cover in place, but it also improves the image quality due to water's refractive index, RI of 1.33.

Liquid paraffin, an oil, is another effective mountant with RI of 1.47, which is very close to the RIs of many textile fibers. Liquid paraffin meets many of the requirements of an effective mountant, including liquids that are colorless, non-swelling, stable, and safe [2]. These authors suggest that because liquid paraffin's RI is 1.47, "Only cellulose diacetate and triacetate fibers of RI approximately 1.46–1.47 are not clearly visible in liquid paraffin and, if their presence is suspected, a second preparation using water or cedar wood oil as the mountant should be made" [2, p. 7]. Other wet mountants used include glycerin (glyceryl) (RI 1.45) and other immersion oils (olive oil RI 1.48 and cedarwood oil 1.513–1.519).

Immersion oil may be applied in two ways: it can be placed on top of the coverslip, keeping the actual specimen from touching the oil, or the specimen can be fully submerged in the oil and then placed on the microscopic slide without the glass slide cover. The oil immersion method, including liquid paraffin, is suitable for fibers so that clearer image can be observed. It works for hair fibers (for example wool) when the anatomy of the fiber needs to be viewed. With the use of immersion oils, one can view the medulla.

4.4 Mounting

When mounting a sample, the fibers should be spread as evenly and parallel as possible in relation to the shorter dimension of the glass slide. To prepare a slide for proper examination, the examiner must check the slide and make sure both the slide and slide cover are clean and free of any impurities. Cross-contamination is very common, especially with novice examiners. If one uses a contaminated slide, the fiber identification may yield incorrect results, as the examiner can easily mistake the contaminant for the fiber. To correctly mount a sample without contamination, the slide must first be cleaned. Then, only a few drops of the mountant should be placed in the center of the slide. Then, the fiber sample should be placed on top of the mountant. If loose fibers are difficult to handle, as they are easily lost, they could be put in a mountant before being placed on the slide. Once on the slide, they may be teased apart with a needle, and then more mountant may be added. The slide cover should then be placed onto the prepared slide. Slide covers must be placed carefully, as they can easily distort the placement of fibers on the slide. Once the slide cover is in place, any excess liquid mountant should be wiped from around the slide cover. The placement of a slide cover should also exclude any air bubbles, as they may be mistaken for fibers during the observation.

5 Fibers

Fibers are the smallest, hair-like parts of a textile fabric. A common definition of a fiber is "a unit of matter characterized by its fineness, flexibility and having a high ratio of length to thickness" [2, p. 1]. When identifying textiles, it is fibers that must be examined under a microscope, as it is the fiber shape that helps to identify the fiber content or fiber type. Some fibers occur naturally in plants (cotton) or in animals (silk); other fibers are artificially made. The four main groups of fibers used in textiles are plant, animal, regenerated manufactured, and synthetic fibers. These will be briefly summarized next and in more detail throughout this workbook (see Figure 3).

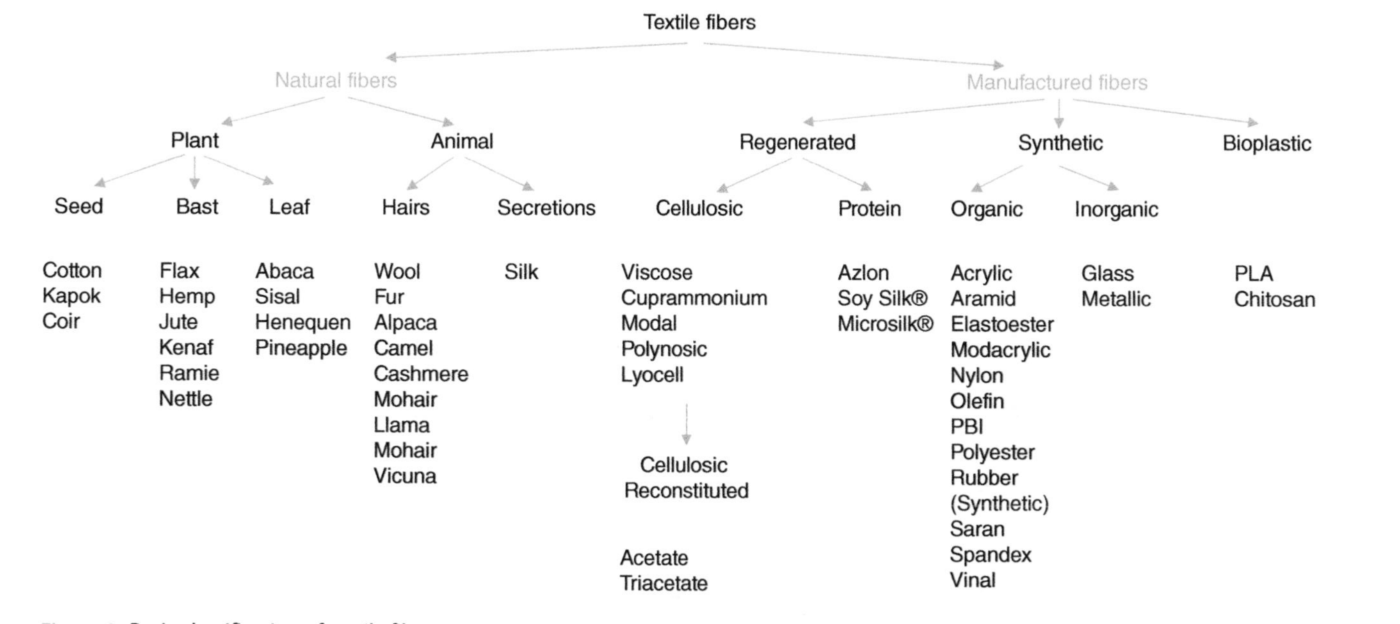

Figure 3 Basic classification of textile fibers.

5.1 Plant Fibers

Fibers from plants are usually referred to as vegetable or cellulosic fibers. The term "cellulosic" is used because all plant fibers are composed of cellulose. Cellulose, found in all plants and trees, is a biomolecule composed of a polysaccharide consisting of chains of many D-glucose units. It is an important structural component of the primary cell wall in plants. Cellulose in plants makes plant fibers strong; however, each plant has a different amount of cellulose. Cotton, for example is composed of 83% of cellulose, whereas linen is composed of 71% of cellulose. Cotton is believed to contain the purest form of cellulose. However, plant-derived cellulose is usually accompanied by other substances such as hemicellulose, lignin, and pectin, amounts of which also determine the strength of cellulosic fibers. Cellulose is a building block of many textile fibers, not only natural plant fibers but also regenerated manufactured fibers (as discussed in the next section). Cellulose fibers under the microscope have an irregular shape and certain characteristics specific to a particular type fiber such as convolutions in cotton. Cellulosic fibers are easily identifiable with the use of microscope.

5.2 Animal Fibers

Animal fibers are mainly animal hair and animal secretions. Just as plant fibers are composed of cellulose, animal fibers are composed of protein molecules. Proteins are biomolecules consisting of chains of amino acid residues. A strong protein and key structural component of hair is called keratin, which gives hair its strength.

Animal hair comes from a variety of animals such as sheep, goats, rabbits. The hair from some animals, such as camels, goats, and rabbits, is called specialty or luxury fibers because it is scarce and harder to obtain than the hair from sheep. The hair from sheep for everyday wool fabrics is simply collected by sheering the animals. Other hair fibers include animal fur.

Animal secretions come from cocoons of the silk moth. These cocoons are collected and unwound to get the fibers out. Animal secretion fibers, for example silk from moth or spider webs, unlike hair fibers have proteins called fibroin. They also have a protein called beta keratin, which is responsible for waterproofing ability of the fiber. As animal fibers have irregular shapes under the microscope, a variety of animal hairs is easily distinguishable under the microscope. For example, the scales on sheep hair have a different shape than the scales on rabbit hair. Therefore, scientists have developed different ways to successfully distinguish animal hair by using the microscope.

5.3 Regenerated Manufactured Fibers

Regenerated fibers are manufactured fibers that are composed of cellulose. This is why they have a feel and exhibit some of the properties of natural fibers such as cotton or silk, i.e. comfort, absorbency, soft to the touch. These fibers are made from naturally occurring polymers in which the cellulose is broken down and reformed into a new matter. Because the cellulose is broken down, regenerated cellulosic fibers do not have the same strength as pure cellulosic fibers. Imitating much desired silk fibers, the first fiber, manufactured this way in 1924, was viscose rayon called “artificial silk” or “viscose.” It had originated in the late nineteenth century as “artificial silk” and later became known as “rayon” in the twentieth century. The name “rayon,” partially developed by the US Department of Commerce, is a combination of “ray” from the sun with “on” from cotton. The reason for this is that the first rayon fibers were bright like a ray of sun and had a hand like that of cotton. Since then there has been an evolution in the manufacture of other regenerated cellulose fibers, such as cuprammonium, high wet modulus (HWM), lyocell, and the development of at least six different types of manufacturing processes. These different rayon fibers were each manufactured to imitate a specific natural fiber, i.e. HWM (modal) was manufactured to imitate cotton.

The cellulose in manufactured regenerated fibers comes from wood also called wood pulp. The wood pulp is usually dissolved into a solution from which rayon fibers are manufactured. The solution is usually pushed through a spinneret with fibers extruded and emerging out the other end. Immediately afterward, the fibers are usually soaked in a solution resulting in fibers that look star-like in cross-section and have striations in a longitudinal view under the microscope. Regenerated cellulosic fibers are of more of a uniform shape because they are all manufactured via the spinneret method. Under the microscope, they are distinguishable from other natural fibers and somewhat distinguishable from other synthetic fibers. There are also regenerated protein fibers, but they are not as common as regenerated cellulosic fibers.

5.4 Synthetic Fibers

Synthetic fibers are petroleum-based fibers, such as polyester and nylon. These fibers are synthesized from petrochemicals. Synthetic fibers are made from a synthetic polymer and are specifically engineered to have certain desirable properties. Because regenerated manufactured fibers, such as rayon, are plant-based, while synthetic fibers are petroleum-based, they are manufactured by different methods – wet spinning vs. melt-spinning. Synthetic fibers, such as nylon and polyester, are chemically manufactured through melt-spinning. The fibers are extruded from a spinneret that determines fiber shape, creating fibers that look very similar. This makes synthetic fibers difficult to distinguish under a microscope. The synthetic fibers discussed in this book include the five major fibers – polyester, nylon, acrylic, spandex, and olefin. The start of the development of synthetic fibers began with nylon. Nylon was the first synthetic fiber, and it was developed by Wallace Carothers, an American researcher who worked at the DuPont chemical company in 1930. It was a perfect time for nylon to debut as the United States was soon to be engaged in World War II. Nylon, a strong fiber, was useful for the manufacture of parachutes and ropes. Nylon replaced silk that was becoming scarce during wartime. It also became popular for women's wear such as stockings. The second synthetic fiber, polyester, came about 10 years after nylon and was introduced by the British chemists, John Rex Whinfield and James Tennant Dickson, in 1941. Because of its easy care, versatility, and affordability, polyester is the most used textile fiber in the apparel industry today.

5.5 Fiber Morphology

Fibers come in a variety of lengths and are always longer than their diameter. Fibers can be of infinite length, and these fibers are called filament, while short-length fibers are called staple. All natural fibers come in the staple form except silk, but manufactured fibers may be made both staple and filament. One of the prerequisites of fibers to be used in textiles is its length. Fibers have to be long enough to be able to be spun into yarn. Shorter fibers are spun with a harder twist so that the fibers will not come apart, while filament fibers, given their infinite length, do not need to be spun with a hard twist. To view fibers under a microscope, fabric yarn has to be untwisted, and fibers have to be pulled out. Because of the hard twist in spun yarns, the fibers are harder to take out from the spun yarn than those from the filament yarn. When preparing a microscopic slide for viewing, a handful of fibers need to be placed on the slide. If you do not separate the fibers well enough and you put too many fibers on the slide, as students often do, you will not be able to identify the fiber characteristics because all you will see is intertwined lines, making it impossible to get a single fiber view. Care must also be taken so that the fibers on a slide do not overlap. At the other end of the spectrum, it is helpful not to have only two or three fibers on the slide because you may have difficult time finding these few fibers, especially with a higher magnification objective such as 100×.

Again, there are two main types of fiber, staple and filament (see Figure 4).

Staple fibers are short, and when spun into a yarn, they make a fuzzy yarn with protruding ends. Yarns made of staple fibers are called spun yarns, and these short fibers must be twisted hard enough to add strength so that they do not come apart.

Filament fibers, on the other hand, are smooth and long, having an infinite length. Yarns made out of filament fibers are called filament yarns. Because of their greater length compared to staple fibers, filament fibers do not need a hard twist to make a yarn. They are slightly twisted, resulting in a smooth yarn in contrast to the fuzzy yarn of staple fibers.

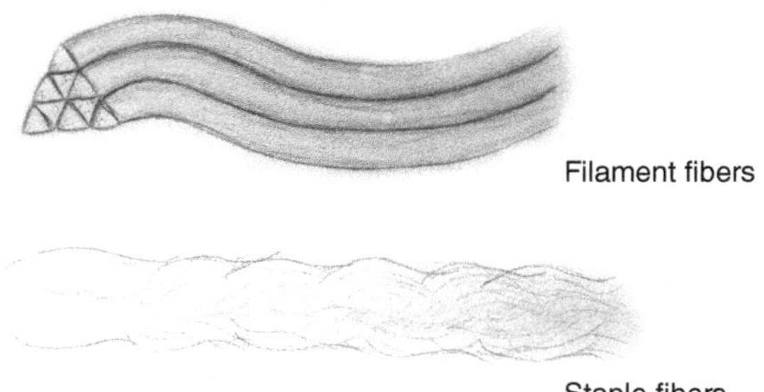

Figure 4 Fiber structure: filament and staple.

5.6 Fiber Shape

A part of a morphological examination of fibers under a microscope is the fiber's longitudinal and cross-sectional view. Fiber's longitudinal (lengthwise) shape is determined by the fiber's cross-section. If a fiber has a round cross-section, it will have a smooth rod-like longitudinal shape. In another example, if a fiber's cross-section is star-like with many ridges, these will create striations (lines) along the length of the fiber (as seen in viscose rayon). Therefore, both cross-section and longitudinal characteristics should be considered when viewing fibers under a microscope. Natural fibers are of different shape and have distinguishing longitudinal and cross-sectional characteristics. However, manufactured fibers do not possess as many distinguishing characteristics as natural fibers do. Many manufactured fibers share similar morphological characteristics, making fiber content difficult to identify.

5.7 Fiber Measurement

As mentioned earlier, a fiber must have a high ratio of length to thickness. The thickness or diameter of a fiber is a very important property because (i) this determines the fineness of fibers which in turn determines the end product made from the fibers, and (ii) it will determine how many fibers will be used to make a yarn. The finer the fiber, the finer the yarn that can be made from it. More fibers in the yarn's cross-section will add strength to the yarn. There are different products made from fine and from coarse fibers [2]. When we refer to fiber size, we also use the term denier. *Denier* is a unit of measure that indicates the fiber thickness and weight. The denier value for microfibers (which are manufactured fibers thinner than silk) is usually less than 1 denier. There are fabrics with low denier count and with high denier count. Fabrics with low denier count are light, sheer, and soft, and fabrics with high denier count are thick, sturdy, and more opaque.

References

1 Goes, F.J. (2013). *The Eye in History*. London: Jaypee Brothers Medical Publishers.
2 Greaves, P.H. and Saville, B.P. (1995). *Microscopy of Textile Fibers*. BIOS Scientific Publishers, Royal Microscopical Society.

1

Natural Cellulosic Fibers

Cellulosic fibers are derived from one of three parts of a plant (i.e. the flower or seed, stem, or leaf) and differ based on their origins. *Seed* fibers are fibers that come from the flower or seed of a plant; *bast* fibers come from the stem of the plant. While bast fibers from different types of plants (e.g., flax or hemp) more often reveal their commonalities under a microscope, there are certain differentiable morphological characteristics that can be identified. For example, the longitudinal characteristics of linen appear as having nodes or kinks, which are similar to that of hemp's longitudinal characteristics; however, their cross-sectional view differ as hemp has a wider lumen than linen.

Fiber morphology is the study of the structure and form of fibers and requires the utilization of microscopes to see the true characteristics of the fiber. When we view the fibers under a microscope, we see either *longitudinal* or *cross-sectional* fiber characteristics (which will both be discussed in this book).

1.1 Seed Fibers

1.1.1 Cotton

Cotton comes from the flower of cotton plant, which is of the genus *Gossypium* in the mallow family Malvaceae. There are few types of the plant. Cotton is an ancient fiber as it has been used for clothing articles for thousands of years. Archeological evidence suggests that cotton was used in Pakistan more than 5000 years ago and in Mexico 3000 years ago, which makes sense because the cotton plant grows in warm climates.

Today, cotton is the most commonly used natural fabric. Its desirable properties such as comfort, absorbency, and good conductor of electricity have made it difficult to replace it with other fibers.

As already mentioned, the longitudinal characteristics of cotton fibers are ribbon-like twists or also often called convolutions (Figure 1.1). These twists can be closer together or further apart depending first on the age of the plant

Textile Fiber Microscopy: A Practical Approach, First Edition. Ivana Markova.
© 2019 John Wiley & Sons Ltd. Published 2019 by John Wiley & Sons Ltd.

Figure 1.1 Longitudinal view of pima cotton fibers featuring ribbon-like twists (2000×).

and second on chemical fiber applications. The cotton fibers that come from a fully ripped bud will have twists closer together, and therefore a little bit more easy to identify under the microscope, than the cotton fibers that come from an half-ripe bud, also called immature fibers [1], which will have twists that are further apart and therefore more difficult to identify under the microscope. When the cotton bud is unripe (immature fibers), the fibers will have very minimal twists or no twists at all. The twist starts forming after the cotton boll opens. Not all cottons have the same amount of twists/convolutions. High-quality long staple cotton has about 300 convolutions/in., and low-quality short staple cotton has no more than 200 convolutions/in. [2].

The advantage of the fiber twist is that it allows the fibers to cling together, and spinning yarns even out of short-length fibers would not be difficult [2]. Also, twists enable cotton fabrics to be more elastic and not stiff like other cellulosic fibers such as linen fabrics. The twists also give cotton its uneven surface which in turn enables only a random skin contact, making the fabric comfortable to wear.

Another reason why these ribbon-like twists could not be easily identified under the microscope is the chemical application called mercerization, which is "a chemical treatment applied to fibers to permanently impart a greater affinity for dyes and various chemical finishes" [3]. The fibers are immersed in a sodium hydroxide solution, which will give cotton fibers greater absorptive properties, higher degree of luster, and higher strength. Mercerization of cotton fibers will change the shape of the fibers because this chemical treatment will cause the fiber to swell, and thus, twists will be flattened out and not easily seen under the microscope (see Figure 1.2).

The cross-section of cotton fibers could also vary in shape based on the maturity of the fiber. In a cotton boll, there are always some immature fibers mixed with mature ones. The mature fibers have a round cross-sectional shape, thick cell wall, and a small lumen (also referred to as central canal). On the other hand, immature fibers do not have round cross-section but have a U shaped cross-section with a thin cell wall (see Figure 1.3). Immature fibers are not desirable because they are more difficult to spin into yarns and to dye [2]. Immature cotton fibers appear collapsed under the microscope. We can explain these differences in more detail by identifying the different parts of the cross-section. There is the outer wall of the fiber, which is also called the primary wall, the inner wall, which is also called the secondary wall, and the center of the fiber cross-section, which is called the lumen (see Figure 1.2). Lumens are the hollow regions in the center of the cross-section. "The reasons for invisible lumens in some cross-sections are that the fibers are either so mature that the inner spaces are fully filled or immature that the openings are totally collapsed" [1, p. 412].

We learn a great deal from knowing about the lengths and widths of fibers as they affect fabric performance and properties. Cotton fibers are generally

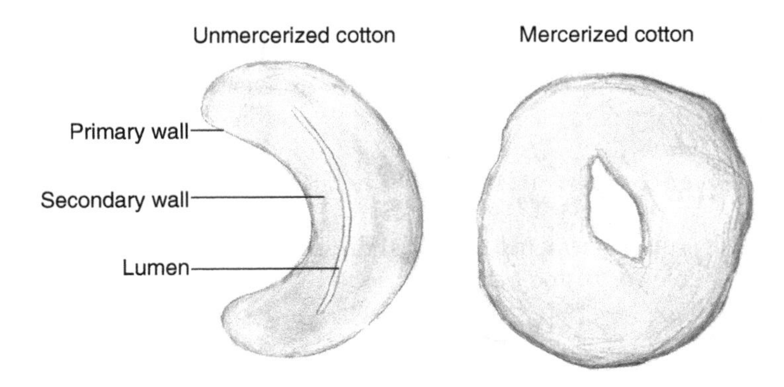

Figure 1.2 Physical structure of cotton fibers.

Figure 1.3 Cross-section of cotton fibers with a variety of wall thickness showing mature fibers with thick cell wall and a small lumen and immature fibers U shaped cross-section with a thin cell wall (1500×).

around 1 in. long; longer cotton fibers generally range from 1½ to 2 in. Cotton fiber length has an effect on fabric properties; the longer the fiber, the better the quality of the fabric. Longer cotton fibers yield a smooth, soft, and lustrous fabric, which is the reason why such fabrics are more expensive and desirable (Figures 1.4 and 1.5). Fabrics made out of long staple cotton fibers are called *Pima, Supima, Egyptian,* or *Sea Island* cottons; cottons with shorter fiber lengths are called *upland* cottons (see Figures 1.6–1.8 for a variety of cotton fibers).

End use: Apparel – undergarments, pajamas, dresses, and children's wear. Interior – curtains, tablecloths, and beddings.

1.1.2 Organic Cotton

Organic cotton is a more sustainable alternative to conventionally grown cotton because it does not use chemicals and pesticides for its growth. Although conventionally grown cotton is a natural fiber, it is not considered to be a

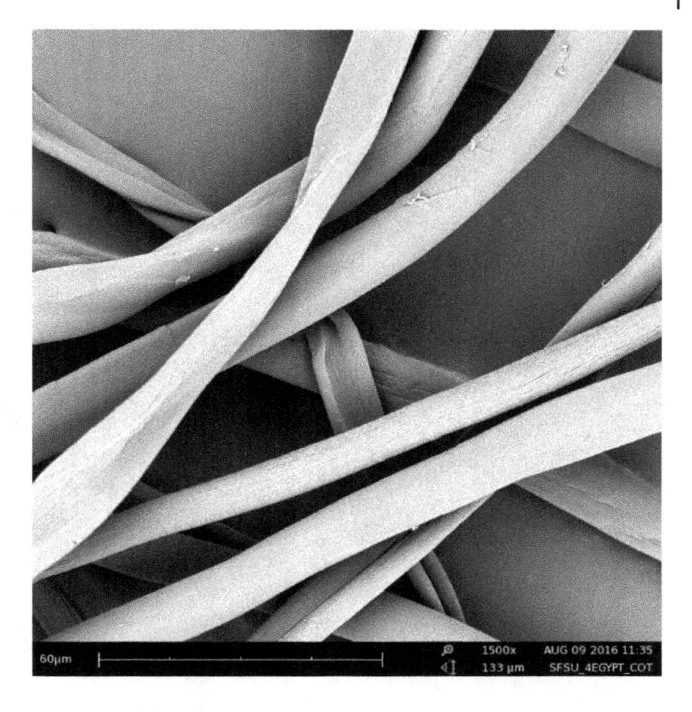

Figure 1.4
Longitudinal view of Egyptian cotton fibers – extra-long fibers of high quality (1500×).

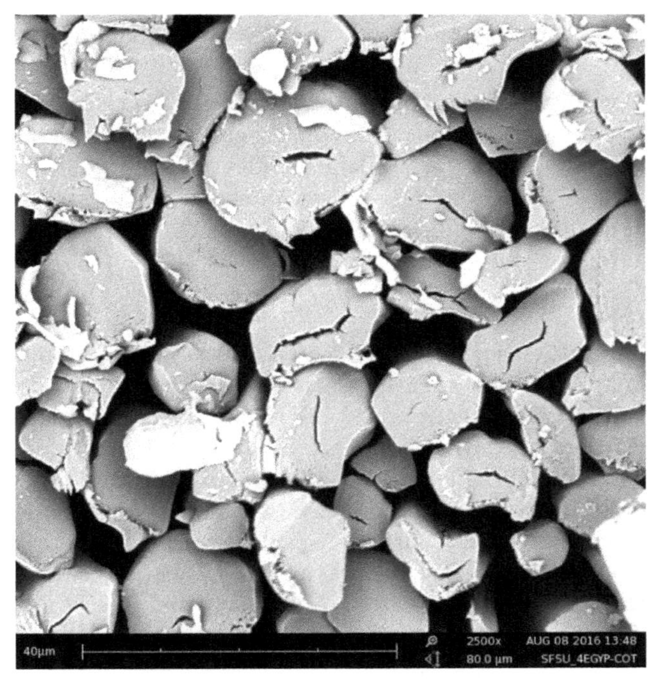

Figure 1.5
Cross-sectional view of Egyptian cotton fibers (2500×).

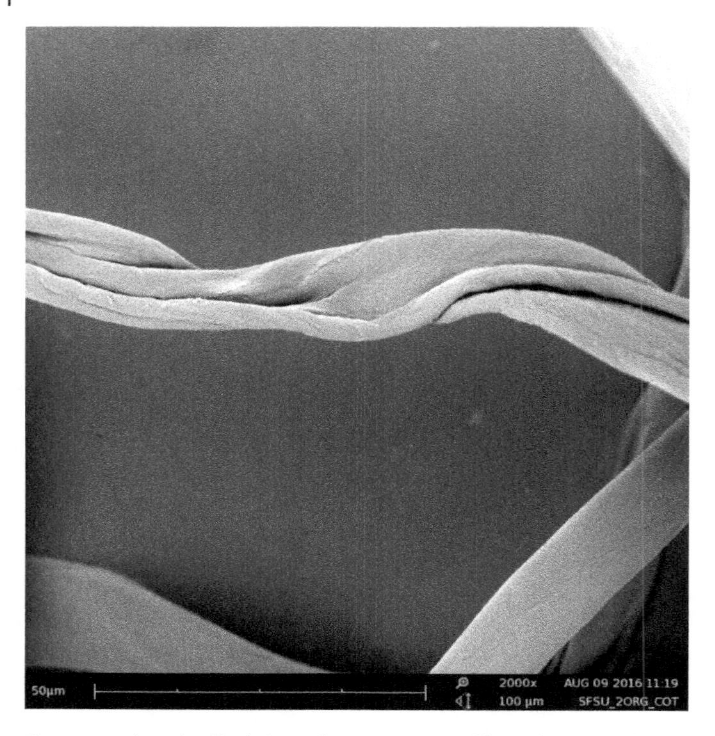

Figure 1.6 Longitudinal view of organic cotton fibers showing ribbon-like twists (2000×).

sustainable fiber because it uses excessive amount of water, chemical fertilizers, and synthetic pesticides during growth. Under a microscope, organic cotton has similar fiber characteristics to conventionally grown cotton (see Figures 1.6, 1.7, and 1.9). The major end uses of organic cotton include wearing apparel, baby clothing, and bedding.

1.1.3 Kapok Fibers

Kapok (Ceiba pentandra) is a seed fiber (e.g. like cotton) derived from the fruit of a tree. Due to their hollowness, the fibers are considered to be the lightest natural fibers in the world [4]. Kapok fibers are shorter than cotton fibers, which negatively affect their strength [5], and their morphology is quite different as well. When viewed under a microscope, the longitudinal characteristics of the fiber resemble microtubes and are not as convoluted as cotton [6]. Their smooth cylindrical surface gives the fiber a natural luster that resembles silk fiber [4]. Also, kapok fiber has a thin wall, which is covered with a thick layer of wax; this makes this fiber nonabsorbent and gives it a hydrophobic property. The waxy surface will immediately repel water, not allowing it to go beyond the

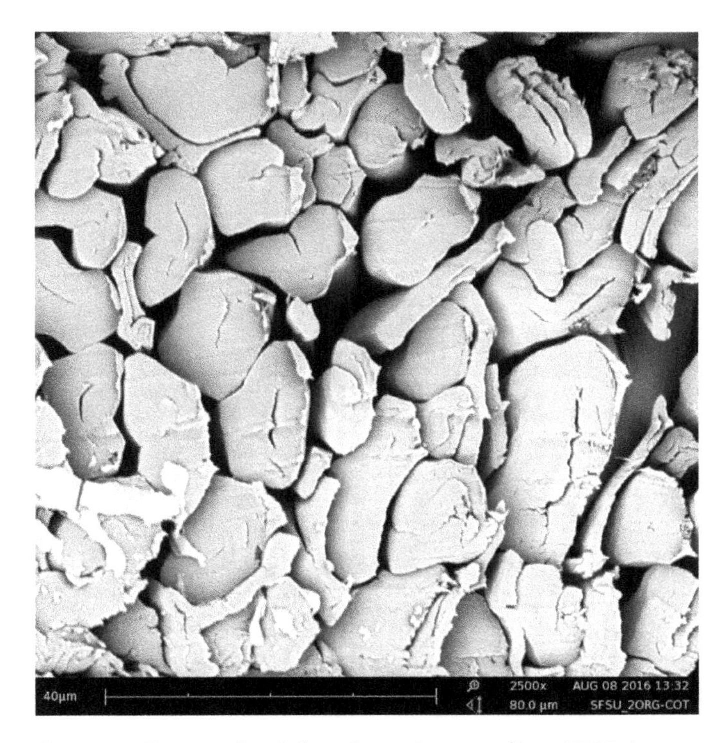

Figure 1.7 Cross-sectional view of organic cotton fibers (2500×).

waxy surface. This is very different from cotton, which is a highly absorbent fiber.

Kapok's nonabsorbent fibers are also difficult to dye, resistant to swelling, and have atypical large hollow lumens, which do not collapse like cotton [7]. In fact, the cross-section of kapok fiber shows a wide open lumen [8], which is oval or sometimes round in shape [7] and is filled with air [4] and large air bubbles [7]. Kapok fibers likely possess the greatest hollowness of all natural fibers [6]. The hollow degree can reach up to 80–90% of hollowness, which allows the kapok fibers to retain warmth and be used as thermal insulators, sound insulators, and water lifesaving materials [9].

Kapok fibers are short (e.g. compared to cotton); they are 10–35 mm long, with diameters of 20–43 mm [7]. Their short length prevents the fibers from being spun into yarns. In fact, to be used for textiles [10], they must be blended with cotton fibers. However, in a kapok/cotton blend, the kapok content should not exceed 50% to yield the best result [11].

Interestingly, although kapok fibers do not absorb water, they do absorb oil [6]. Thus, the wetting liquid for kapok is oil, and the nonwetting liquid is water [12]. This oil-absorbent property is of interest to ecologists and

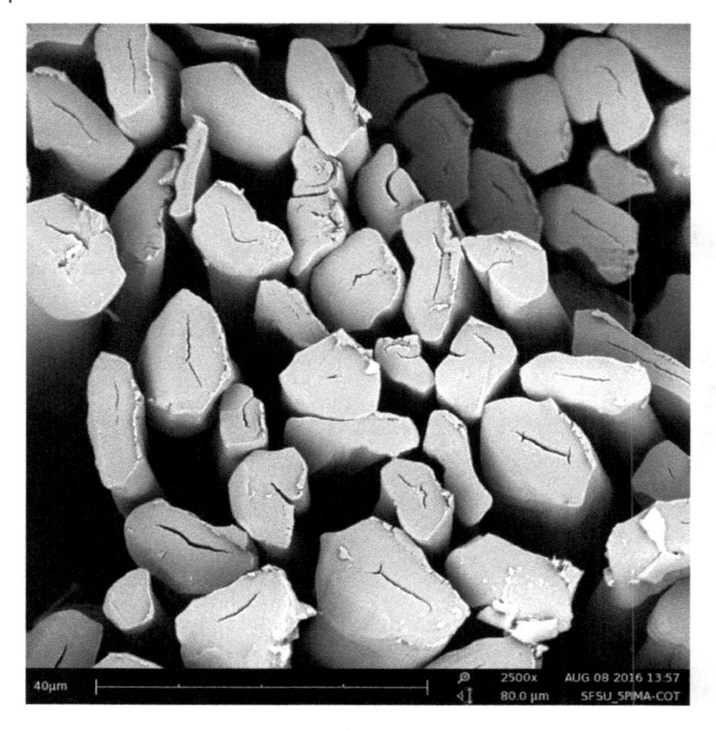

Figure 1.8 Cross-sectional view of pima cotton fibers – long staple fibers (2500×).

environmentalists because the material can be of use during oil-spill cleanups. Researchers attribute this oil-absorbent property to the fiber's micro-nano-binary structure [13]. On the surface of the fiber, there are two coats of protective layering so that water droplets do not penetrate the fiber wall; these layers are on micro and nano levels.

End use: Kapok fiber is light, very buoyant, resilient, and resistant to water. It is used as filler in pillows, mattresses, insulation, and soft toys.

1.1.4 Poplar Fibers

Poplar fibers (*Populus*) are another kind of seed fiber; they were brought to the public's attention in the last decade by German scientists who announced it as the fiber of the year in 2006 due to the sustainable way it is grown. Researchers are still trying to figure out the best way to utilize this fiber in textiles; one of the problems with poplar fibers is that they are too short to be spun into yarns. The structure of poplar fibers is similar to kapok; however, they are shorter. A longitudinal view of the fibers reveals microtubes with smooth cylindrical shapes. The cross-section of the fiber is round with a thin wall and a large hollow lumen.

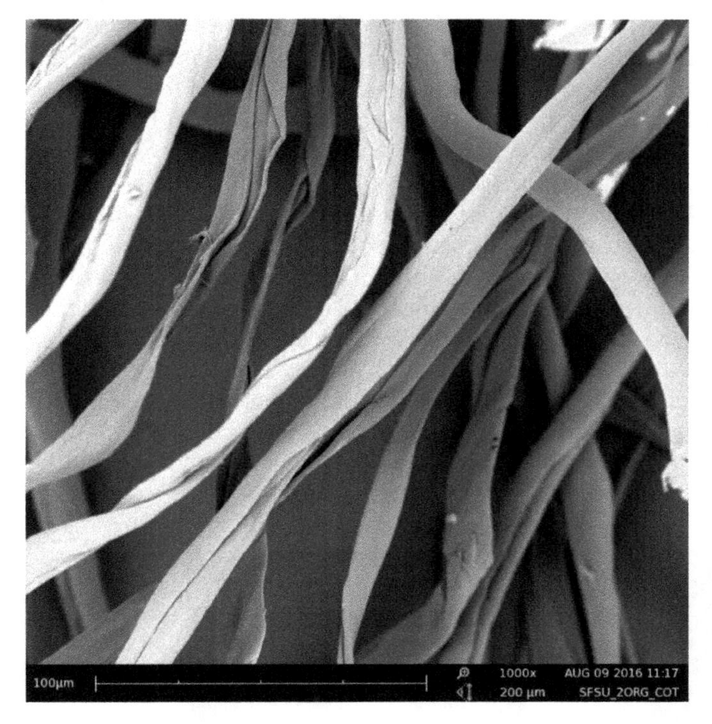

Figure 1.9 Longitudinal view of organic cotton fibers depicting similar fiber characteristics (ribbon-like twists) to conventionally grown cotton (2000×).

Even though poplar (vs. kapok) fibers are small, their hollow spaces are relatively large (i.e. 89% vs. kapok's 77%) [12]. Thus, the density of poplar (vs. kapok) fibers is lower. Poplar fibers also possess hydrophobic ability (i.e. they repel water but absorb oil). Thus, it would be difficult to apply dye stuff on poplar seed fibers; the fibers are therefore mainly used as insulation or stuffing materials. In fact, technology has not yet developed a way to utilize poplar fibers in the clothing industry. Similar to kapok fibers, poplar fibers may be blended with cotton fibers to make them suitable for spinning. Because of poplar fiber's hydrophobic property, the fiber becomes a good oil-absorbent material [14]. The advantage of poplar seed fibers is found in their short length when it comes to oil absorbency. The shorter the fiber, the better the oil absorption capacity. Poplar fibers are around 0.95–1.59 in. in length and 8–14 μm in diameter.

1.1.5 Willow Fibers

Willow (*Salix*) fibers are seed fibers that come from the same tree family as poplar fibers. One of the advantages of willow (vs. poplar) fibers is the ease of harvest; also, these fibers are considered fibers of the future because they are ecofriendly and sustainable. The environmental benefits are found in their cultivation and growth as they do not require significant irrigation (e.g. as cotton does); also, the willow tree can help remove contaminants from the air, soil, and water surrounding it. Willow fibers have similar characteristics to kapok and poplar fibers. A longitudinal view reveals a microtube with a smooth, cylindrical shape; specifically, the fibers appear smooth and silk-like with some ribbon-like twists. The cross-section of the fiber appears round with a thin wall and a large hollow lumen (see Figures 1.10 and 1.11). Even though willow fibers are similar in structure to those of kapok and poplar, they are shorter and thus would be difficult to spin into yarns. Willow fibers are around 0.4–0.6 cm in length and 4–15 μm in diameter [15].

Potential future uses: These new ecofriendly/sustainable fibers can be utilized in many ways in textiles (e.g. in manufacturing fiber composites). They can also be used as insulators in jackets, pillows, or stuff toys and can be blended with other fibers.

Figure 1.10 Longitudinal view of willow seed fibers depicting microtube with smooth, cylindrical, and ribbon-like shape (1000×).

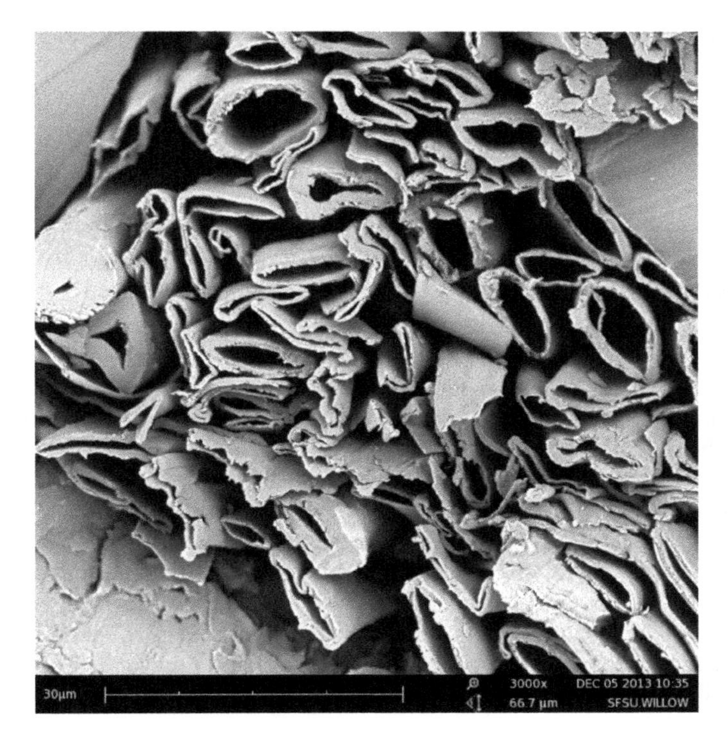

Figure 1.11 Cross-sectional view of willow fibers showing round, thin wall with large hollow lumen (3000×).

1.1.6 Coir Fibers

Coir fibers come from the outer shell of coconuts, are coarser than the other seed fibers previously discussed, and are grouped into one of two categories. The first group is extracted from mature coconuts, and it is brown in color and coarser. The second type is extracted from immature coconuts; the fiber is finer and white in color. However, overall, coil fibers are inherently stiff and ductile. The fibers have a rough, textured surface under a microscope. Their most distinctive characteristic is a highly porous surface. However, the cross-section of coir fibers is somewhat similar to kapok, poplar, and willow fibers because of the significant hollowness of the fibers. Coir fiber has a large, hollow, tube shape with very thin walls. The size of coir fibers also makes them quite unique; they are much larger than other seed fibers. Their length reaches up to 35 cm, and they extend 11–25 μm in diameter.

End uses: Interiors – upholstery, rugs, doormats, mattresses, sacks, and insulation. It is also used for geotextiles.

1.2 Bast Fibers

Bast fibers such as flax, nettle, ramie, hemp, and jute have very similar morphological characteristics and thus are difficult to distinguish under the microscope. Distinguishing among these fibers is not done easily, but some researchers propose certain methods that help to point out some differences.

Bast fibers, whether soft or hard, are easily distinguished from other cellulosic seed fibers or protein fibers. Bast fibers have a distinguishing characteristic in longitudinal view, many of them have cross-marks (single oblique lines) on the fiber surface. These cross-marks are "cell dislocations which are usually accentuated by the mechanical treatment during processing to extract the fiber from the plant" [16, p. 123]. The cross-sectional view usually (for most but not all bast fibers) entails a polygonal shape and lumen (which is an opening cavity in the center of the fiber).

The fibers' cross-section shape would be the only way to distinguish some of the bast fibers. For example, there are similarities between flax, hemp, and jute fibers' cross-sectional shape as they have a rounded polygonal outer shape with an oval or round narrow lumen. Ramie and nettle fibers' cross-sectional shape differs slightly from the aforementioned fibers. The latter have a more elongated not as polygonal and round cross-sectional outer shape and have a larger lumen [17]. Bast fibers such as nettle, hemp, and jute are important to identify mainly when dealing with archeological finds, as today they are not used in large quantities in clothing.

1.2.1 Linen

Linen (Linum usitatissimum) is a widely used fiber, comes from the stem of the flax plant, is also referred to as flax, and has a harsher texture than that of seed fibers (e.g. cotton). Bast fibers have many similar properties to other cellulosic fibers (e.g. absorbency, good electrical conductivity, and comfort); however, bast fibers have a harsher hand than seed fibers. This property is reflected in the fiber shape, which we can see under a microscope. Linen is an ancient fiber, and there is plenty of archeological evidence that it was used for clothing for thousands of years in Egypt and in Europe.

Linen fibers have some variations in shape based on their source and quality; however, they also share some common characteristics. Under a microscope, longitudinally, linen fibers resemble bamboo sticks. Their bamboo-like joints (also called nodes) are key characteristics of linen fibers, distinguish them from seed fibers (e.g. cotton), and are cylindrical in shape (see Figures 1.12–1.14). These nodes add to the flexibility of the fibers (which bend at the nodes) and, of course, add to the texture of the fabric. In some cases, the nodes intersect like the letter X [18].

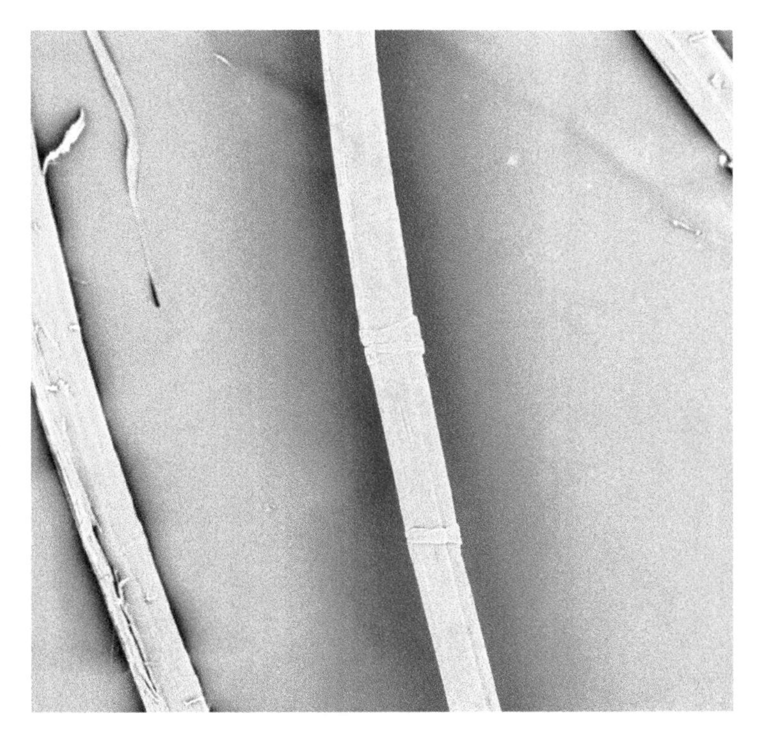

Figure 1.12 Longitudinal view of linen fibers showing cross-marking or nodes – resembling bamboo-like joints (1000×).

The cross-sectional shape of linen fibers is very different than that of seed fibers (e.g. cotton) as it is made up of irregular polygonal shapes. The fiber may have several polygonal sides with a small lumen, which is the center void portion of the fiber. "The cross-sectional contour of flax is sharp-edged, polygonal, slightly elongated. The lumen is visible as a small round to oval opening in the center" [19, p. 27]. The cell wall of linen fibers is very thick, and thus, the lumen looks like a narrow line; this significantly differs from seed fibers' hollow, tube-like, and cross-sectional shapes (see Figure 1.15). It is common to see fiber bundles when viewing cross-sectional shape of bast fibers under a microscope (see Figure 1.16). The size of linen fibers is much greater than that of seed fibers (e.g. cotton). The individual fibers are as long as 10–12 in. in length [20]; the approximate diameter is 15–17 μm.

End use: Apparel – suits, dresses, and blouses. Home interiors – draperies, upholstery, tablecloths, and dish towels.

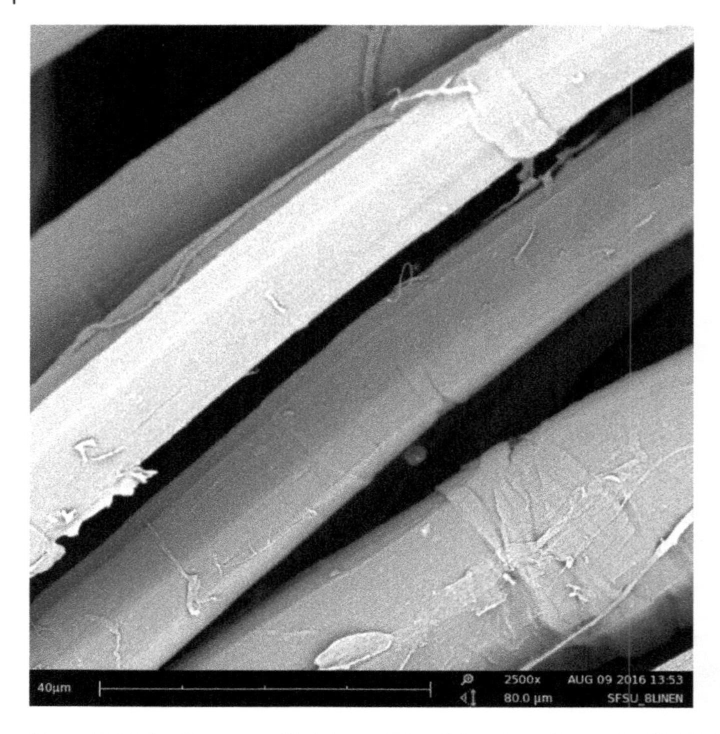

Figure 1.13 Further magnified view of linen's bamboo-like nodes (2500×).

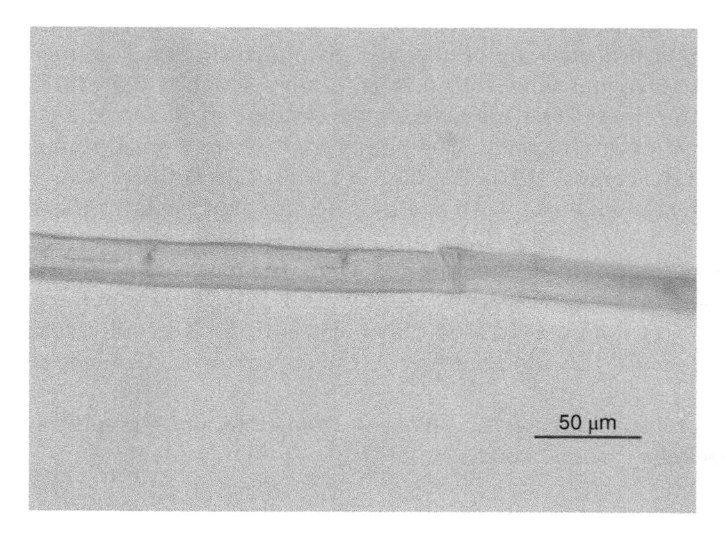

Figure 1.14 Linen fiber showing bamboo-like nodes through a compound microscope.

Figure 1.15 Cross-sectional view of linen individual fibers (ultimates) and fiber bundles. Individual linen fibers showing polygonal fiber shape (2000×).

Figure 1.16 Depiction of fiber bundles composed of individual fibers of polygonal shape. Bast fibers come in fiber bundles, which can be separated into fiber ultimates.

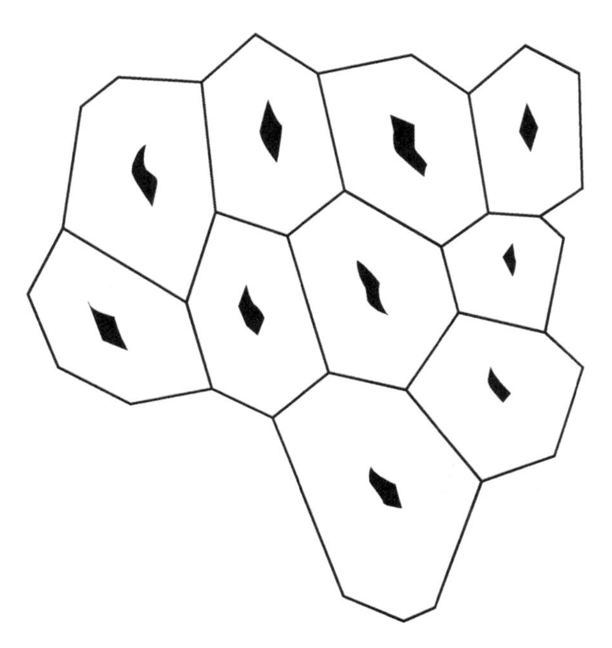

1.2.2 Ramie

Ramie (Boehmeria nivea), a bast fiber, which is a part of the Urticaceae (Nettle) family, is also known as China grass. Ramie plant is native to Southeast Asia (Philippines), China, Japan, and Brazil because warm and humid is the best climate for its growth. Ramie fibers have a great strength compatible to synthetic fibers [21]. Ramie has some properties of those of linen such as low elasticity, absorbency, stiffness, low resiliency, and high natural luster. Similar to flax, it breaks with frequent folding. End uses: Ramie is often blended with other fibers and used in blouses, shirts, and suits. Ramie is also utilized in home furnishings such as table cloths, pillows, and window treatments. Because of its strength, ramie is used for ropes, industrial sewing threads, and fishing nets. Polarized light microscopy (PLM) is suitable for anisotropic fibers such as these bast fibers. Besides the oblique cross-section markings, long and striated longitudinal lines (see Figure 1.17) are visible in bast fibers such as ramie. The distinguishing longitudinal characteristics of ramie fibers are pronounced diagonal cracks [19] (see Figure 1.18). Cross-markings or bamboo-like nodes are also seen in ramie,

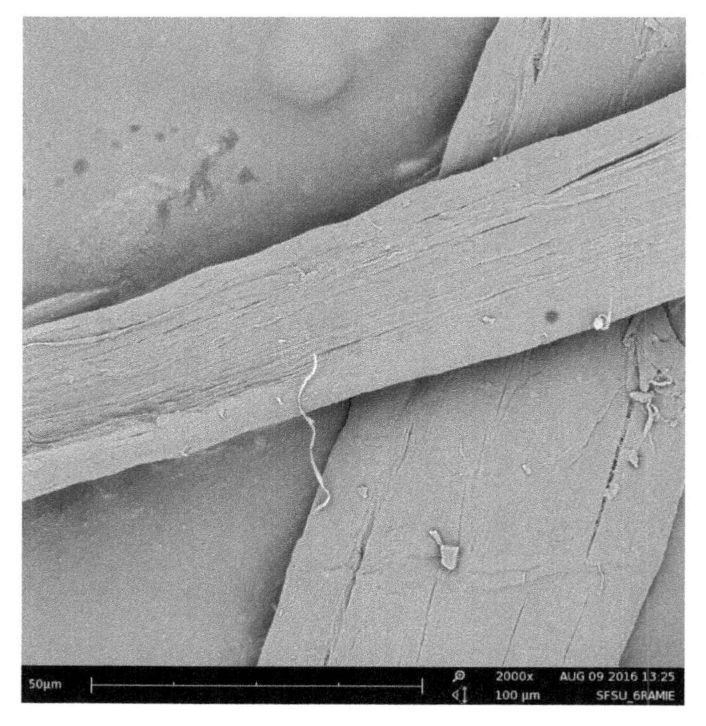

Figure 1.17 Further magnified ramie fibers showing fibrillary orientation typical of bast fibers (2000×).

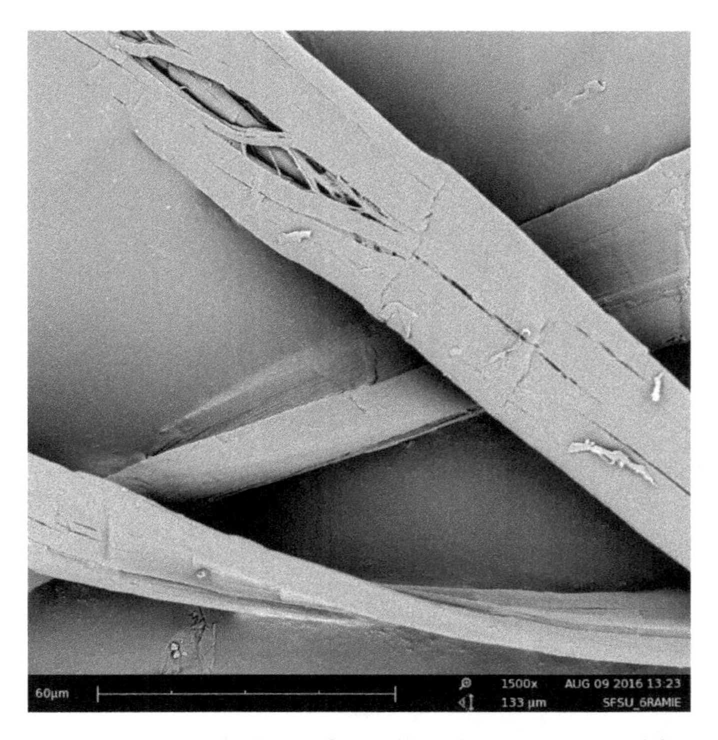

Figure 1.18 Longitudinal view of ramie fibers showing pronounced diagonal cracks (1500×).

but mostly under polarized microscope [19] (see Figure 1.19). Cross-sectional shape of ramie is similar to cotton, but it has a larger diameter [19]. The shape also varied from hexagonal to oval [19] (see Figure 1.20).

1.2.3 Hemp

Hemp plants (Cannabis sativa) grow in the northern hemisphere and include many different varieties of this plant species. Hemp fibers are coarser than flax fibers, and therefore, they are great to use for ropes [22]. Hemp fibers viewed under a transmitted light will show the longitudinal view nodes or kinks (or also called fiber dislocations). Also, when hemp fibers are viewed with scanning electron microscopy (SEM), fibers will show nodes (or dislocations) very similar to linen. See Figure 1.21. However, when magnification is increased above 1000×, the nodes are less visible, and bark-like lines (striations) could be identified (similar to jute) (see Figure 1.22). Therefore, higher magnification does not necessarily give better results. The cross-sectional view of hemp fibers are believed to be similar to those of flax and jute as they have a polygonal shape and a lumen in the middle. Some cross-sectional distinguishing

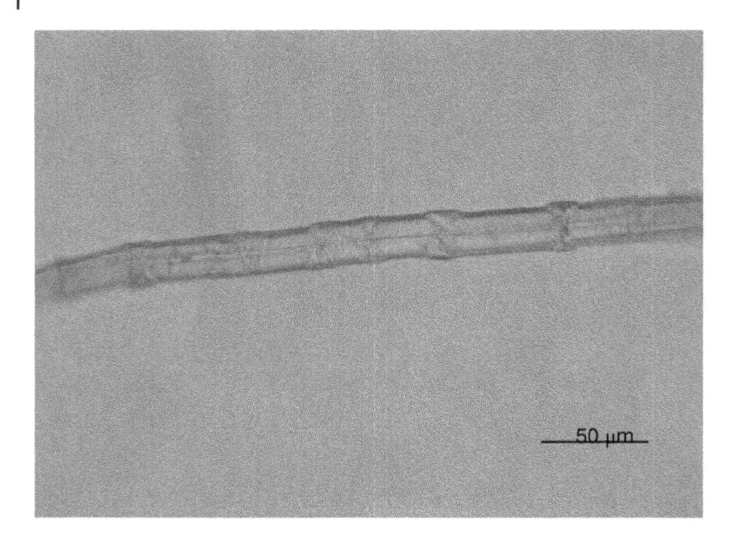

Figure 1.19 Longitudinal view of ramie fibers depicting cross-markings typical to bast fibers and viewed under compound microscope.

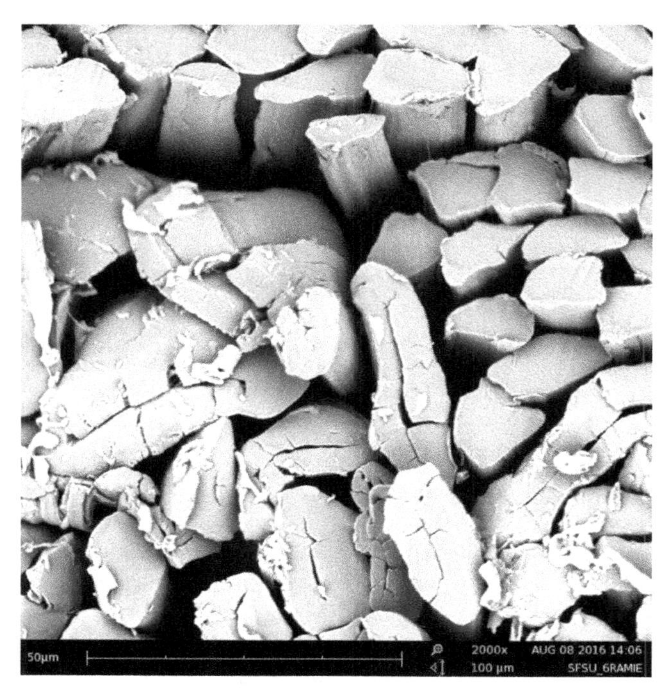

Figure 1.20 Cross-section of ramie fibers depicting a variety of features ranging from features found in cotton's cross-section, hexagonal, and oval (2000×).

Figure 1.21
Longitudinal view of hemp fibers showing bamboo-like nodes typical for bast fibers (1000×).

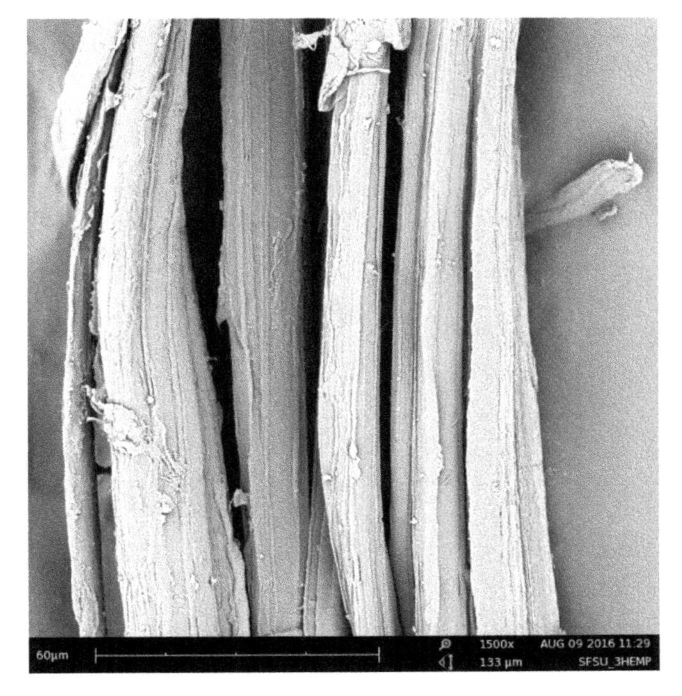

Figure 1.22
With increased magnification (1500×), bamboo-like nodes are becoming less visible and bark-like lines (striations) are more prominent in hemp fibers.

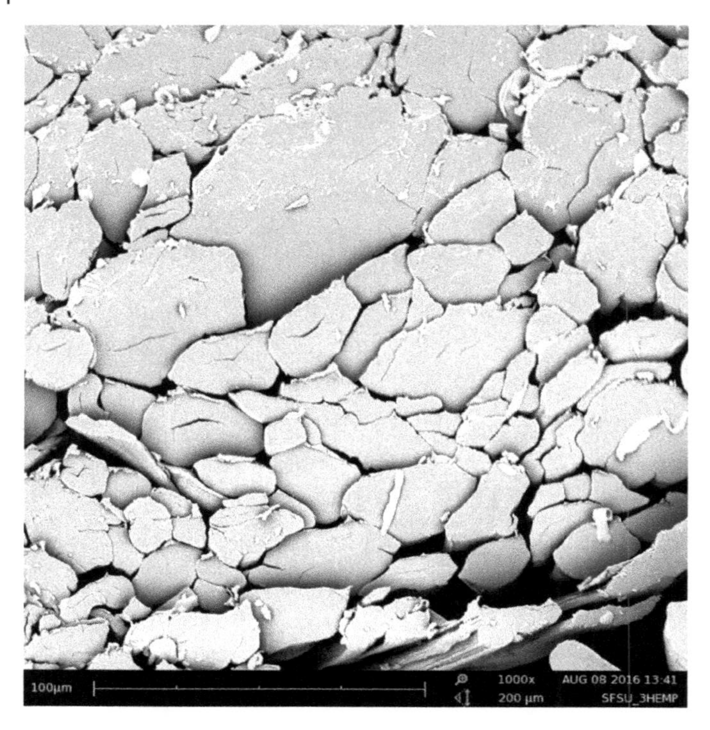

Figure 1.23 Cross-sectional view of hemp fibers depicting polygonal shape and a lumen in the middle (1000×).

characteristics from flax are that its corners are rounder, and the lumen appears as an elongated cavity in the middle [19] (see Figure 1.23). However, hemp has a wider lumen than flax [19], and its diameter is much larger in comparison to flax and jute (see Table 1.1). When viewing a single fiber via PLM, a thicker line

Table 1.1 Fiber diameter of bast fibers.

Fiber	Diameter (µm)	References
Flax	12–20	[23, 24]
Bamboo	6–12	[23]
	5–20	[25]
Hemp	32–34	[26]
	18–23	[19]
Nettle	19–47	[27]
	19–25	[28]
Jute	15–20	[19]
Ramie	30–70	[19]

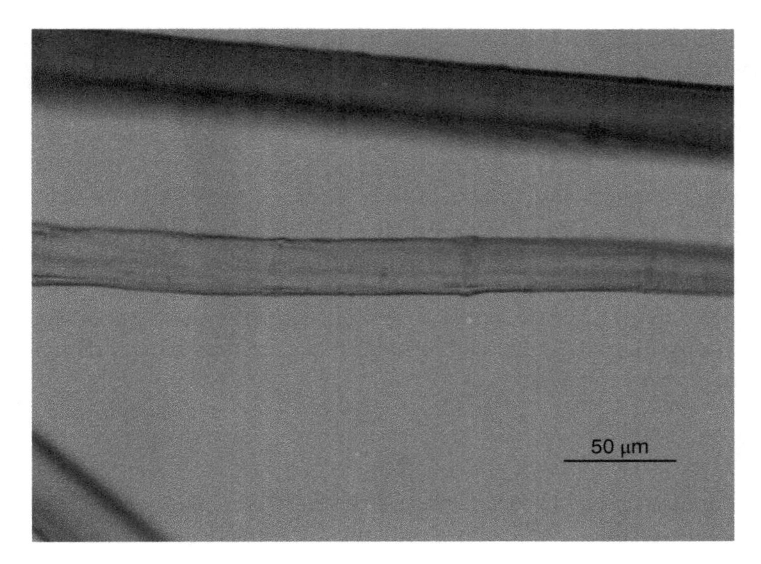

Figure 1.24 Longitudinal shape of hemp fibers using polarized microscope, which allows us to see a thicker line (striation) running down the middle of the fiber – lumen.

(striation) running down the middle of the fiber could be seen – this line would be the lumen (a cavity of the bast fibers) (see Figure 1.24).

1.2.4 Bamboo

Bamboo fibers have been used since ancient times especially in Asia. It has been used for many different purposes in Asian history such as papermaking and construction. The shape of bamboo fiber is smaller in comparison with a flax fiber [23]. The spindle-like short fibers are only around 1.9 mm in length and 15.3 μm in diameter, which have a tapered shape at both lengths. The fibers are densely packed within the fiber bundles, and their diameter is around 5–20 μm (mean of 13.16) [25]. Sizing of fibers could also be a distinguishing feature among bast fibers as bamboo fibers are much smaller in size from those of flax, hemp, and nettle fibers. Bamboo fibers are only 2 mm in length, and therefore, there are two ways in which bamboo fibers can be prepared for the textile industry. First, because of its short length, bamboo in its fiber bundle form can be processed chemically and physically treated [23]. Second, bamboo (is retted into bamboo pulp) fiber pulp is used to make bamboo regenerated fibers. The fibers from the latter method would have different morphological characteristics from the first method, as the latter method is similar to viscose rayon.

The longitudinal characteristics of bamboo fibers are similar to those of other bast fibers; however, two main fiber surface differences are present. First of these distinguishing characteristics is that when viewing a single bamboo fiber under the microscope, it is rough (appears as having tree-bark stripes) in its surface (rougher than flax or ramie). The second distinguishing characteristic is that bamboo fibers lack the nodes or cross marks, which are seen in flax or ramie fibers [23].

Bamboo fibers have more differences in their cross-sectional view as they have a round shape with a small round lumen, compared to polygonal shape (small lumen) of flax and ellipse shape (with lumen and cracks) of ramie [23]. Bamboo's cross-sectional shape and size could distinguish it from other bast fibers; although the shape of single bamboo fiber is far smaller than that of flax [23]. As a result, single bamboo fiber can be identified according to the characteristics of cross-section and size.

1.2.5 Jute

Jute fiber is yet another bast fiber that comes from the *Corchorus* genus plant family, which grows in India, Bangladesh, or China. Jute fibers are not as long as other bast fibers and therefore are difficult to spin. In addition to its inadequate length, jute fibers are also weaker than other bast fibers. Jute fibers are weak when wet and when exposed to sunlight and therefore are not used for as much for ropes and fishnets as other bast fibers. End use includes twine and cordage, but it is also made into a burlap fabric that is usually used for sacking, carpet backing, or in gardening and geotextiles. Jute fibers are coarse and therefore not commonly used in clothing products. Under a microscope, jute fibers share similar characteristics with other bast fibers as they have polygonal to oval cross-sectional fiber shape [19]. However, nodes and cross-marking are found only infrequently [19] (see Figures 1.25–1.29).

1.2.6 Fiber Size

One of the identifying characteristics can be the width of the fiber or the diameter of the fiber. Bast fibers vary in fiber diameters. Fiber diameter could be determined by the use of an ocular micrometer that could be attached to a microscope (see Table 1.1).

1.2.7 Nettle

Nettle is a part of the Urticaceae family and belongs to the genus *Urtica*. Stinging nettle and ramie are a part of the same family. The stinging nettle is the most common species grown in Europe. Nettle, flax, and hemp fibers have been extensively used in Europe in the past before the cotton industry took over. It was also a very important fiber in prehistory for Native Americans. Nettle is considered a biodegradable fiber as it comes from a renewable source

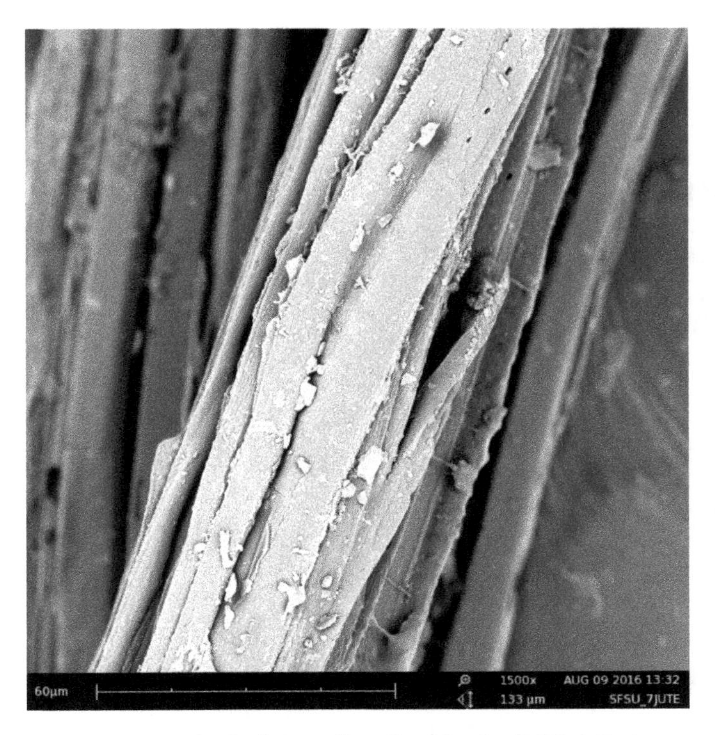

Figure 1.25 Longitudinal view of jute fiber bundles (1500×). Nodes are rarely seen in jute fibers.

and requires little water and energy for production. Nettle plant grows like weed, and it does not need extra work for growth. The longitudinal view of nettle fibers is nodes or kinks (cross-marks), again characteristics as seen in hemp and flax. The cross-section of nettle has a polygonal shape but more of an elongated band formed outer shape and a large lumen; this shape is similar to ramie.

Because some bast fibers have similar longitudinal and cross-sectional characteristics, they are difficult to identify under a microscope. New identifying methods have been utilized in identifying bast fibers, also with the use of microscopes. Two methods that will be discussed next are identifying the fibrillary orientation of bast fibers and the presence of crystals in bast fibers [17]. Both of these methods have distinguishing features and for best results should be used together. The first method, identifying fiber fibrillary orientation, is conducted with the use of PLM. An examiner focuses on the direction of the fibrillary cell of the bast fibers. These fibrils are basically bundles of microfibrils, which are positioned in a helical pattern along the length of the fiber [17]. The orientation of the helical pattern can go either to the left or to

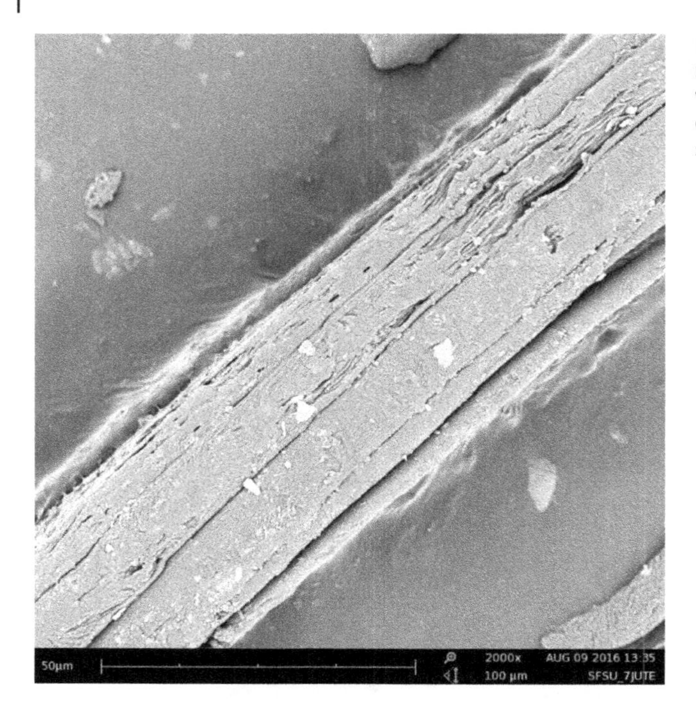

Figure 1.26 Further magnified (2000×) view of jute fibers depicting bark-like surface texture.

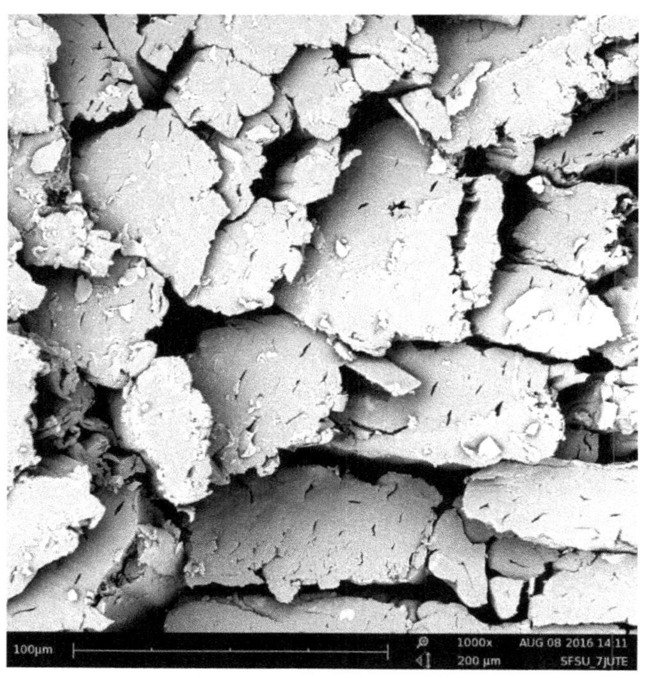

Figure 1.27 Cross-section of jute fibers (1000×). Fibers are mostly found in bundles.

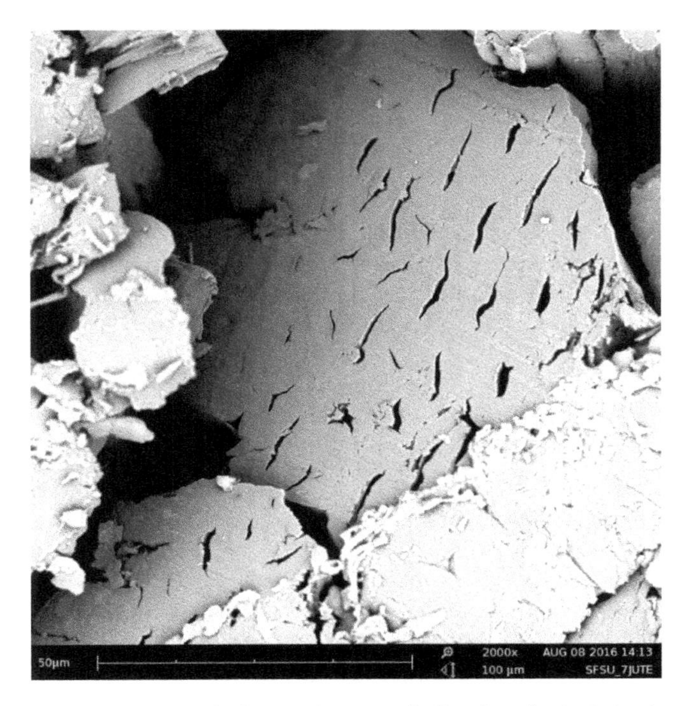

Figure 1.28 Magnified (2000×) image of a fiber bundle depicting jute fibers having large lumen opening in the center of each fiber.

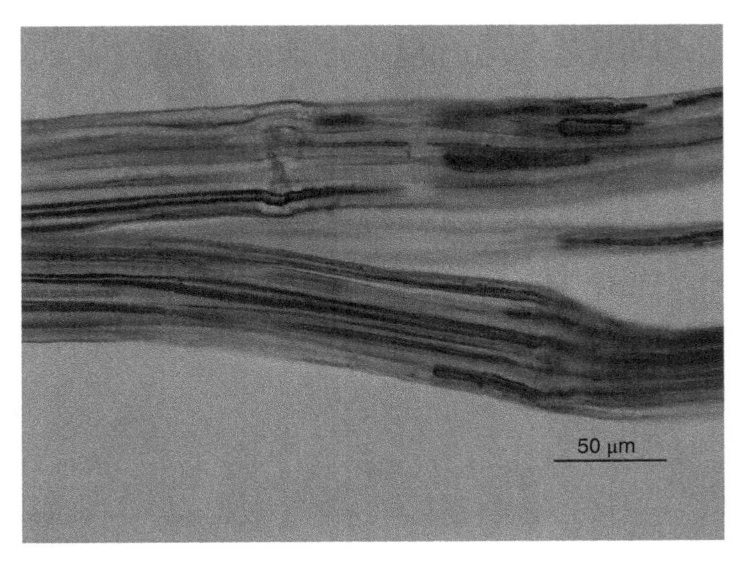

Figure 1.29 Longitudinal view of jute fibers using compound microscope – nodes are more visible than with the use of SEM.

the right. In the case that the fibrils are oriented to the right (or also referred to as having the S-twist), the fiber is flax, nettle, or ramie, and if the orientation of the fibrils is to the left (or also referred to as having a Z-twist), the fibers are hemp or jute [17].

To further distinguish, for example flax and nettle (both of which would have the S-twist fibril orientation), a crystal test is recommended. For this test also, PLM is recommended with a mounting medium – galvanol. During this crystal test, an examiner would be able to detect calcium druses (cluster of crystals) or solitary crystals. Calcium druses occur in nettle, ramie, hemp, and jute, and solitary crystals occur in hemp and jute. Flax fibers have no crystals present.

1.2.8 Bast Fiber in Its Historical Context

As mentioned earlier, bast fibers such as nettle, jute, and hemp are not used today in clothing on a large scale. However, the identification of them becomes very important in historical clothing. The earliest textiles are not only believed to come from plants such as flax but also from plants with wild origins [22]. As a matter of fact, there is more and more evidence found that nettle fibers were used for clothing. In some instances, historians believed archeological items were made of flax (because the characteristics of bast fibers are so similar), but later rediscovered that those fibers were actually made out of nettle. This misidentification occurs every so often, but with new methods for identifying textile fibers such as before mentioned fiber fibrillar orientation or detection of crystals, we can more precisely understand the use of nettle fibers. One instance is the Bronze Age 2800-year-old Lusehoj burial site in Voldofte, Denmark, where fabric (which was wrapped around a deceased body) was identified as nettle, but previously thought of as flax [29]. The correct fiber identification helps us understand the importance of nettle in clothing thousands of years ago. Because the buried body was wrapped in nettle fiber, it gives the historian an idea that nettle fibers must have been special in some way and might have been considered a luxury fiber [29]. In another similar instance of a Norwegian Viking burial ship that dated back approximately 834 BCE, a cloth from flax plant was found at this site, which was later re-examined and proven to be made out of nettle plant [22].

1.3 Leaf Fibers

Leaf fibers come from plant leaves and leaf stalks. Although they are not used in textile products to the same extent as seed and bast fibers, they can still be found in many interior design products such as rugs and home decor. Leaf fibers are stiffer and coarser than bast and seed fibers. There are three main leaf fibers: sisal, henequen, and abaca. Sisal and henequen fibers are both derived from Agave plants. The Agave family also hosts a plant that is used for making tequila.

1.3.1 Sisal

Sisal fibers are derived from the leaves of Agave sisalana plant, which originates in Mexico. The Aztecs and Mayans used sisal fibers for garments and used the spines of the plant as needles, thus came the name for sisal as "needle and thread plant" [30, p. 204]. The coarse texture of sisal fibers constitutes their use mainly used for rope, baskets, and nets. The longitudinal view of sisal fibers is somewhat cylindrical [7], but it does not show any distinguishing features [19]. The longitudinal view has some indentations, and some fibrils can be seen on its surface. The cross-section of sisal fibers is more pronounced, and the fibers have distinguishing characteristics than those in their longitudinal view. The fibers have polygonal shape with a pronounced circular to oval lumen [19]. Sisal fibers, just like coir fibers, have a large empty lumen and a thin wall. The fibers come in bundles. The size of the fiber can be a good fiber indicator as sisal fibers are larger than other seen or bast fibers. The diameter of sisal fibers is between 100 and 300 μm, and the length ranges between 0.6 and 1.5 mm [31].

1.3.2 Henequen

Henequen fibers are obtained from the Henequen plant (Agave fourcroydes), which is in the same plant family as sisal. Henequen is also an ancient fiber of the Mayans and is native to Yucatan, Mexico. Henequen fibers are similar to those of sisal longitudinally. In cross-section, the shape varies as it changes from bean-like to rounded [7]. Henequen fibers are also used for robes and twine; however, they are coarser than sisal fibers and therefore not considered as of high quality as sisal.

1.3.3 Abaca

Abaca fibers are taken from the leaves of banana-like plant called Musa textilis. The origin of the plant is the Philippines and therefore is sometimes referred to as Manila hemp. The longitudinal view shows both fibrils and some cross-markings. Although cross-markings are believed to be rare [7]. The cross-sectional view also has a large lumen as sisal fibers, but abaca does not have as thin a wall as sisal fibers. Abaca fibers in cross-section are round and not polygonal as sisal fibers. Abaca fibers also come in bundles. When comparing the bundles of abaca and sisal fibers, sisal fibers have bundles of crescent form, and abaca fibers have bundles of oval contour [19]. Abaca is considered to be the strongest of all natural fibers [7]. It is resistant to water, and therefore, it is used for marine ropes and tea bags. The fiber diameter and length of abaca fibers are smaller than those of sisal as the diameter ranges from 14 to 50 μm and the length ranges from 2.5 to 13 mm [32].

1.3.4 Pineapple Leaves

A novelty leaf fiber worth noting comes from pineapple leaves grown in the Philippines. It is considered to be sustainable because the pineapple leaves are a natural waste product. It has been marketed under a Piñatex™ and offered as a sustainable alternative to leather. The fabrication is a nonwoven textile material coated with polylactic acid (PLA) for the feel and appearance of leather-like material. PLA is also considered to be an environment friendly material because it is derived from renewable agricultural sources such as corn starch or sugarcane. It is a synthetic biodegradable polymer. Piñatex fabrication promises to replace other nonenvironmentally friendly alternative such as polyvinyl chloride. Preliminary microscopic examination of pineapple leaves is that in the longitudinal view some indentations and some fibrils can be seen on its surface.

References

1 Huang, Y. and Xu, B. (2002). Image analysis for cotton fibers. Part I: longitudinal measurements. *Textile Research Journal* 72 (8): 713–720.
2 Kadolph, S.J. and Langford, A.L. (2002). *Textiles*, 9e. Pearson Education: Upper Saddle River, NJ.
3 Encyclopedia Britannica. Mercerization. Textile technology. https://www.britannica.com/technology/mercerization (accessed 18 July 2018).
4 Mani, G.K., Rayappan, J.B.B., and Bisoyi, D.K. (2012). Synthesis and characterization of kapok fibers and its composites. *Journal of Applied Sciences* 12 (16): 1661–1665.
5 Fengel, D. (1986). Studies on kapok. *Holzforschung* 6: 325–330.
6 Zhang, Y. and Wang, A. (2014). Kapok fiber: structure and properties. In: *Biomass and Bioenergy: Processing and Properties* (ed. K.R. Hakeem et al.), 101–110. Switzerland: Springer International Publishing.
7 Smole, M.S., Hribernik, S., Kleinschek, K.S., and Kreze, T. (2013). Plant fibers for textile and technical applications. In: *Agricultural and Biological Sciences: Advances in Agrophysical Research* (ed. S. Grundas and A. Stepniewski), 369–397. Intech Publishing.
8 Mwaikambo, L.Y. and Bisanda, E.T.N. (1999). The performance of cotton-kapok fabric polyester composites. *Polymer Testing* 18: 181–198.
9 Liu, X., Yan, X., and Zhang, H. (2015). Sound absorption model of kapok-based fiber nonwoven fabrics. *Textile Research Journal* 85 (9): 969–979.

10 Liu, J. and Wang, F. (2011). Influence of mercerization on micro-structure and properties of kapok blended yarns with different blending ratios. *Journal of Engineered Fibers and Fabrics* 6 (3): 63–67.

11 Yang, L., Bi, S.M., and Hong, J. (2013). Effect of blending ratio on kapok fiber cotton blended yarn property (in Chinese). *Cotton Textile Technology* 41: 30–32.

12 Lim, T.T. and Huang, X.F. (2007). Evaluation of hydrophobicity/oleophilicity of kapok and its performance in oily water filtration: comparison of raw and solvent-treated fibers. *Industrial Crops and Products* 26: 125–134.

13 Zhang, X., Fu, W., Duan, C. et al. (2013). Superhydrophobicity determines the buoyancy performance of kapok fibers aggregates. *Applied Surface Science* 266: 225–229.

14 Likon, M., Remskar, M., Ducman, V., and Svegl, F. (2013). Populus seed fibers as a natural source for production of oil super absorbents. *Journal of Environmental Management* 114: 158–167.

15 Markova, I. (2014). Willow fibers: Environmental solutions. Poster presented at the American Association of Family and Consumer Sciences (AAFCS) Western Region Biennial Conference in Burlingame, California (29 March 2014).

16 Schaffer, E. (1981). Fiber identification in ethnological textile artifacts. *Studies in Conservation* 26 (3): 119–129.

17 Bergfjord, C. and Holst, B. (2010). A procedure for identifying textile bast fibers using microscopy: flax, nettle/ramie, hemp and jute. *Ultramicroscopy* 110: 1192–1197.

18 Woolman, M.S. and McGowan, E.B. (1921). *Textiles: A Handbook for the Student and the Consumer*. New York, NY: MacMillan Company.

19 Von Bergen, W. and Krauss, W. (1942). *Textile Fiber Atlas: A Collection of Photomicrographs of Common Textile Fibers*. New York, NY: American wool Handbook Company, Barnes Printing Company.

20 Watson, K. H. (2007). *Textile and Clothing*. ebook Produced by Stan Goodman, Karen Dalrymple, and the Online. Distributed Proofreading Team at www.pgdp.net.

21 Goda, K., Sreekala, M.S., Gomes, A. et al. (2006). Improvement of plant based natural fibers for toughening green composites – effect of load application during mercerization of ramie fibers. *Composites Part A Applied Science and Manufacturing* 37: 2213–2220.

22 Barber, E.J.W. (1991). *Prehistoric Textiles*. Princeton, NJ: Princeton University Press.

23 Yueping, W., Ge, W., Haitao, C. et al. (2010). Structures of bamboo fiber for textiles. *Textile Research Journal* 80 (4): 334–343.

24 Baley, C. (2002). Analysis of the flax fibers tensile behavior and analysis of the tensile stiffness increase. *Composites Part A Applied Science and Manufacturing* 33: 939–948.

25 Zhou, H. and Zhong, W. (2003). Development and application of bamboo fiber. *Wool Textile Journal* 4: 30–36.

26 Kundu, B.C. (1942). The anatomy of two indian fiber plants, *Cannabis* and *Corchorus* with special reference to fiber distribution and development. *Journal of Indian Botanical Society* 23: 93–129.

27 Bacci, L., Baronti, S., Predieri, S., and Di Virgilio, N. (2009). Fiber yield and quality of fiber nettle (*Urtica dioica* L.) cultivated in Italy. *Industrial Crops and Products* 29 (2–3): 480–484.

28 Bodros, E. and Baley, C. (2008). Study of the tensile properties of stinging nettle fibers (*Urtica dioica*). *Materials Letters* 62: 2143–2145.

29 Bergfjord, C., Mannering, U., Frei, K.M. et al. (2012). Nettle as a distinct bronze age textile plant. *Scientific Reports* 2: 664.

30 Schmidt, B.M. and Klaser Chang, D.M. (2017). *Ethnobotany: A Phytochemical Perspective*. Wiley.

31 Mohanty, A.K., Manjusri, M., and Drzal, L.T. (2005). *Natural Fibres, Biopolymers and Biocomposites*. Boca Raton, FL: CRC Press, Taylor & Francis Group.

32 Hearle, J.W.S. and Peters, R.H. (1963). *Fiber Structure*. London: The Textile Institute, Butterworths.

2

Animal Fibers

2.1 Wool

Wool is an ancient fiber and one of the earliest fibers to be spun into yarn. Wool comes from the hair of sheep, probably the first animal to be domesticated [1]. The fleece of a wild sheep, or primitive sheep, prior to being domesticated, consisted of a long outercoat (kemp) and a light downy undercoat. The fleece of today's sheep is mainly the soft undercoat [1]. Although kemp fibers still exist in various breeds of sheep, different wool types have been developed without the kemp fibers, such as the Merino wool (which will be discussed later).

Wool is a fiber with many valuable properties, and when combined, no other synthetic fiber is their equal. These include good moisture absorption, the ability to be shaped by heat and moisture, excellent heat retention, feltability, and flame retardancy [1].

Wool is not simply hydrophilic, similar to many plant fibers, nor hydrophobic, like many synthetic fibers. Wool fibers have tiny scales on their surface adding a new dimension to the properties of wool. The scales do not allow water droplets to enter the fiber as they repel water while still absorbing water vapor. Thus, wool fibers absorb moisture without feeling wet. This type of fiber, called hydroscopic, is the reason why protein fibers are comfortable to wear [1]. The most recognizable part of wool fibers are its scales on the outside of the fiber. These scales have their own particular functions; among them is the scales' ability not only to repel water but also to repel dirt. The scales also help reduce the wear of the fiber.

Although different breeds of sheep produce wool with different characteristics, for consumers, garments are labeled simply as wool. This designation as wool may include not only hair from sheep but also hair from animals such as goats (Angora and Cashmere goats) and camelids (llama, alpaca, camel, and vicuna) [1].

Textile Fiber Microscopy: A Practical Approach, First Edition. Ivana Markova.
© 2019 John Wiley & Sons Ltd. Published 2019 by John Wiley & Sons Ltd.

When viewing wool under a microscope, there is no need for a high magnification as the scale pattern can be successfully viewed under lower magnification. When viewing the medulla, a lower magnification is also recommended. Immersion oil must be applied when viewing the medulla under a light microscope. In identifying fibers, it is important to note that scale patterns are instrumental in differentiating animal fibers such as cashmere and wool. However, scale patterns cannot be used in differentiating animal hairs from animals that are closely related [2]. Identifying wool from different breeds of sheep is a difficult task [2].

All of these fiber types have three major anatomical regions that are very important in the classification and identification of hair fibers. These are the cuticle, the outer layer containing scales that are also called the epidermis, the cortex, the main body, and the medulla, the central canal [3, 4] (see Figure 2.1).

2.1.1 Cuticle

The cuticle consists of a nonfibrous layer of scales and of an epicuticle, which is a thin membrane covering the scales [1]. Wool fibers can be easily distinguished from other natural and synthetic fibers by merely viewing the overlapping scales (cuticle) on the surface of the fiber as this is the unique structure of wool fibers (see Figure 2.2). No other fibers other than animal hair fibers have these scales. However, when scientists want to see differences between sheep's wool fibers and specialty fibers, they further classify the scales based on the scale pattern, e.g. the arrangement and configuration of scales. There are five different scale pattern types: mosaic, coronal,

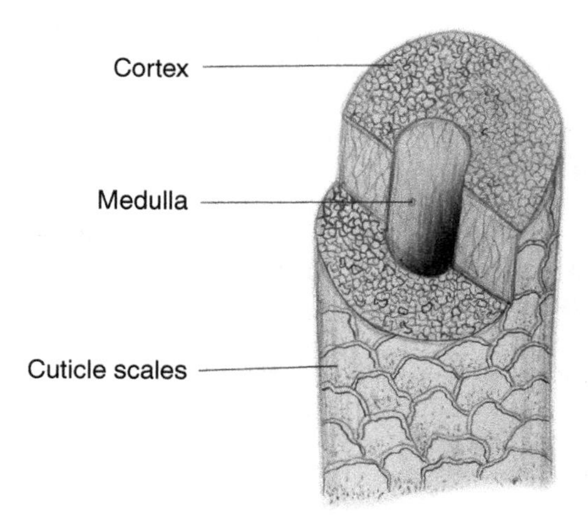

Figure 2.1 Fiber structure of wool.

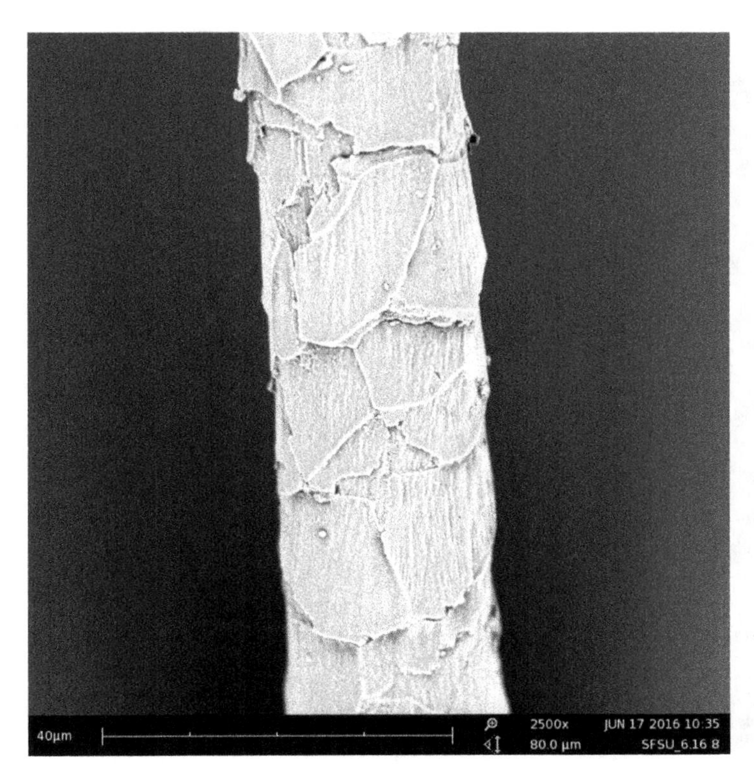

Figure 2.2 Magnified (2500×) sheep's wool fiber showing overlapping cuticle scale structure comparable to shingles on a roof.

pectinate, chevron, and petal [2]. Fibers have a variety of scale patterns. For example, in fine fibers, the cuticle scales completely encircle the fiber, and normally, each scale overlaps the bottom of a preceding scale [1]. The scale arrangement forms smooth fiber margins. Because of this scale arrangement, the scales do not stick out of the fiber, which makes the fibers smoother on the skin and nonirritating. On the other hand, coarse or medium coarse fibers have small scales shaped like fish scales and are dense on the fiber [1] (see Figure 2.3). The scanning electron microscope (SEM) is more precise in capturing the cuticle scale than the light microscope; therefore, majority of micrographs are taken via the SEM. SEM's magnification and depth of focus surpass that of light microscope [5].

2.1.2 Scale Pattern Type (Animal Hair)

Terms describing the different types of scale patterns [2] will be presented next (see Figure 2.4).

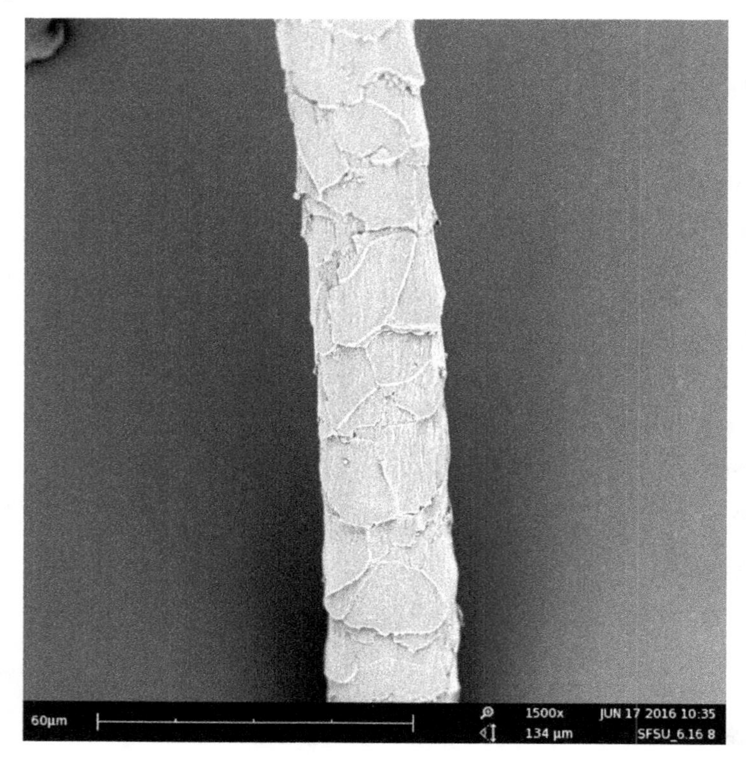

Figure 2.3 Longitudinal view of sheep's wool showing irregularity in sheep's wool (1500×).

2.1.2.1 Mosaic

A mosaic pattern is divided into two categories, regular mosaic and irregular mosaic.

Regular mosaic describes a pattern that is composed of sections that are of the same size.
Irregular mosaic describes a pattern that has sections of differing sizes.

2.1.2.2 Wave

A wave scale pattern is waved. There is a wide variety of wavy patterns: simple regular, interrupted regular, interrupted irregular wave, and streaked wave. In addition, wave patterns are further described as being shallow, deep, and medium.

2.1.2.3 Chevron

A chevron pattern is described as waves with a deep V-shaped pattern. These waves may be regular or irregular and may have crests at the top or at the bottom of the troughs.

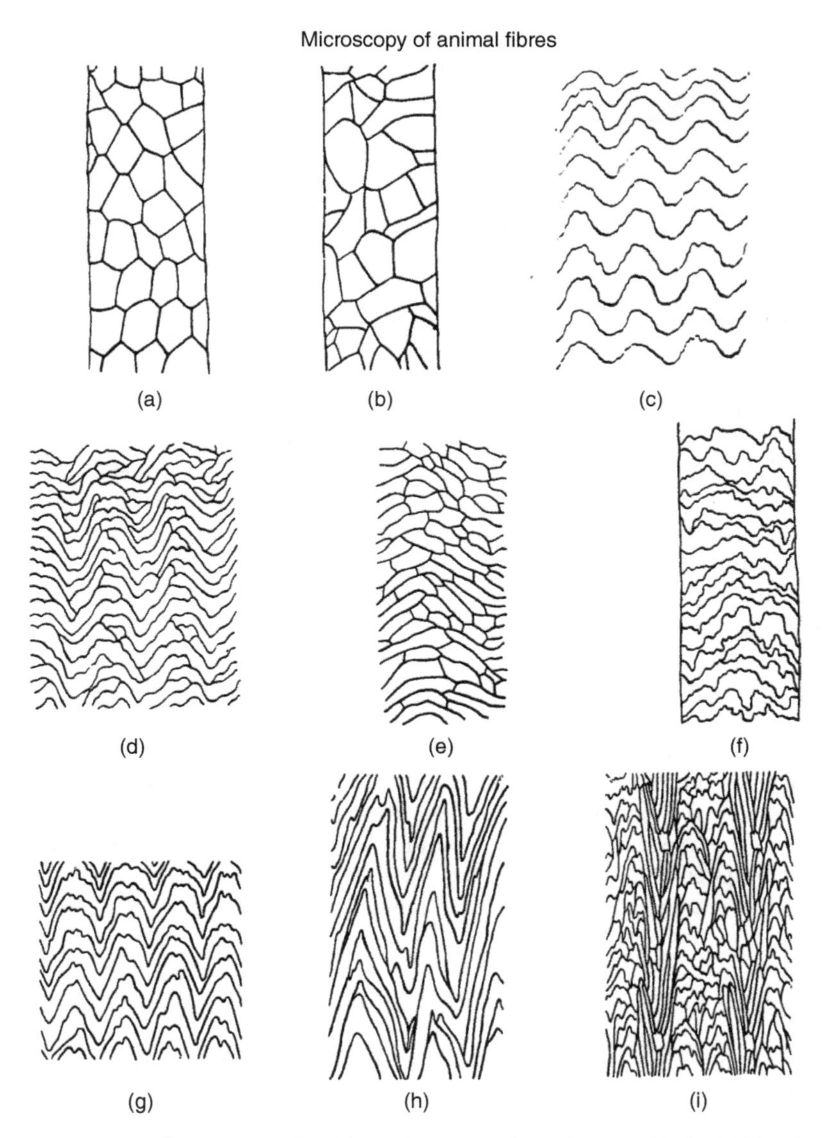

Figure 2.4 Different types of cuticle-scale patterns found in animal hairs. (a) Mosaic (regular); (b) mosaic (irregular); (c) simple regular wave; (d) interrupted regular wave; (e) mosaic (irregular waved); (f) wave (medium depth); (h) single chevron (a form of regular wave); (h) double chevron; (i) streaked wave (a variety of interrupted wave). *Source:* Wildman 1954 [2].

2.1.2.4 Petal

A petal pattern is described as overlapping flower petals. Different petal types may also be noted, but these may be best seen if rolled impressions are created.

2.1.3 Types of Scale Margins

Fiber experts further distinguish fibers by the edges of the scales, referred to as the form of the scale margin. These edges are not always smooth and may be of differing characteristics such as crenate, rippled, or scalloped [2].

Smooth margins are those with margin edges that have a straight-line. Crenate margins have shallow indentations in which there are sharply pointed teeth. Rippled margins also have indentations, as do crenate margins, but are deeper. The teeth or peaks in a margin may be of sharp or rounded points. Scalloped margins have a slight wave on its edges. We do not see these as often in textile fibers.

To successfully view cuticle scale patterns, a rolled impression is recommended because it allows one to see the entirety of a wave scale pattern. Rolled impressions are impressions taken of the entire cuticle surface surrounding the fiber. A rolled impression is not a cast as a cast is taking an impression of only a part of the surface of the fiber [2].

2.1.4 Cortex

The cortex is the core of the fiber surrounding the medulla if a medulla is present. It is made out of cells, which are tapered, long, and flattened [1]. These cortical cells are responsible for wool's unique crimp property as the cells on both sides of the fiber twist and turn around its axis. The cells in the cortex are not the same on each side. As they differ, each side of the fiber behaves differently creating an irregular waviness. Crimp helps wool fibers to cling together and strengthens the yarn [1].

2.1.5 Medulla

Another classification parameter showing the differences between animal hair fibers is the medulla. The medulla is found within the cortical layer, running down the center of the fiber. It is the central canal of a wool fiber in which the color or pigment is carried and which provides air space [6]. The size and shape of medullae varies greatly. It may consist of a continuous, interrupted, or fragmented line [7]. The medulla is mainly seen in coarse and medium wool fibers. More specifically, review of older literature concluded that the medulla is only identifiable in coarse wool fibers with diameters that are larger than 35 μm [8]. Fine wool does not have a visible medulla.

The medulla contains air spaces that provide the thermal, insulating, and light weight properties in wool [8], and when seen under a microscope, it has a dark appearance because it is filled with the mounting medium. When a slide is prepared, to view the medulla, students can use immersion oil and not water droplets. The immersion oil fills up the empty air space of a medulla and therefore appears dark when viewed under a light microscope. The presence of medulla fibers seems whiter than those without a medulla. Light microscopy is recommended to view the medulla, not the SEM (which is recommended for scale formation viewing). The basic types of medulla will be discussed next [2] (see Figure 2.5).

2.1.5.1 Lattice

The lattice type medulla is very wide in proportion to the total width of the fiber [2]. This ornamental looking network of bars of keratin outlines the polyhedral-shaped spaces [2]. Because the medulla is wide and the gases within the medulla occupy a large amount of the medulla, the fiber appears dark when viewed under transmitted light. However, when viewed under reflected light, the medulla appears silvery [2]. This lattice type of medulla can be seen in coarse sheep fibers, kemp fibers, and in the fibers of other animals such as red deer or reindeer [2].

2.1.5.2 Simple Unbroken

The unbroken medulla appears as the simple continuous central canal of a fiber. It is not as wide as the previously mentioned medulla lattice but comes in a range of widths, from wide to quite narrow. Again, when viewed under transmitted light, it appears dark. This type of medulla can be seen in alpaca or calf fibers.

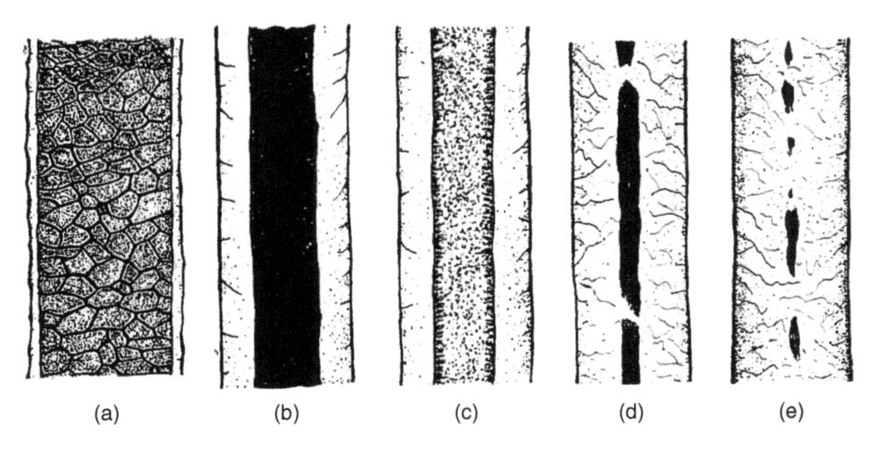

Figure 2.5 Different types of medullae found in animal hair fibers. (a) Unbroken (wide) lattice (usual appearance when not infilled by mounting medium). (b) Simple unbroken. Medium width (usual appearance). (c) Simple unbroken. Medium width (appearance when infilled by mounting medium). (d) Interrupted. (e) Fragmental. *Source:* Wildman 1954 [2].

2.1.5.3 Interrupted

The interrupted medulla is broken at intervals, which can be very irregular. For the most part, it is narrow though it may vary in width. This type of medulla may be seen in medium-quality sheep fleece fibers.

2.1.5.4 Fragmental

The fragmental medulla could be described as vanishing, though it irregularly reappears. At times, the fiber could be identified as unmedullated. It consists of tiny fragments running down the center of a fiber.

2.1.5.5 Ladder Type of Medulla

The ladder-type medulla consists of a series of patches running down the center of a fiber. These patches are evenly spaced out, and the fiber, when viewed under a transmitted light microscope, which darkens the patches, appears like a ladder. There are two types of ladder medulla: the *uniserial*, with only one ladder running down the fiber longitudinally, and the *multiserial*, with multiple ladders running down the fiber. Sometimes a fiber has a variation of uniserial and multiserial medulla in one fiber. Ladder medullas also range in sizes from narrow to quite wide. While ladder medulla is not seen in sheep fleece fibers, it is observed in fur fibers such as rabbit, muskrat, and cat (see Figure 2.6).

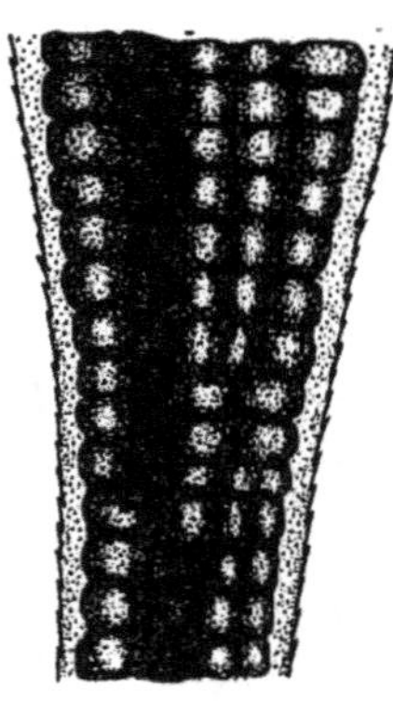

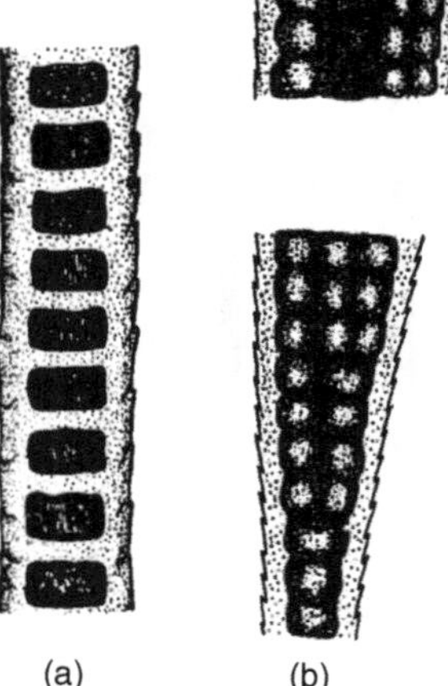

Figure 2.6 Ladder type of medulla seen in rabbits. (a) Uniserial and (b) multiserial. *Source:* Wildman 1954 [2].

2.1.6 Fiber Size

To further differentiate animal hair fibers, the diameter or width of the fibers also varies greatly. The fiber diameter, also called fineness, is an important determinant of fiber quality and fiber type. A small fiber diameter produces fine fibers that are found in soft, smooth, and luxurious fabrics such as cashmere or fine wool. Wool or coarse wool have a larger fiber diameter, averaging around 33–44 μm, than those found in fine or cashmere wool averaging 14–17 μm [9]. The length of wool fibers, although not frequently used as a distinguishing characteristic, also varies based on the breed of sheep it comes from. Generally, the length ranges from 1 to 14 in. or longer. Interestingly, fine wool fibers are usually shorter, and coarse wool fibers are longer [6].

2.1.7 Fiber Morphology

Under a microscope, coarse wool or wool from sheep fibers are roughly cylindrical, irregular, and have a rough surface. The scales are responsible for the surface properties of the fiber. In sheep's wool, the scales do not lie flat/close to the body/cortex but point outwardly (see Figure 2.7). The scales'

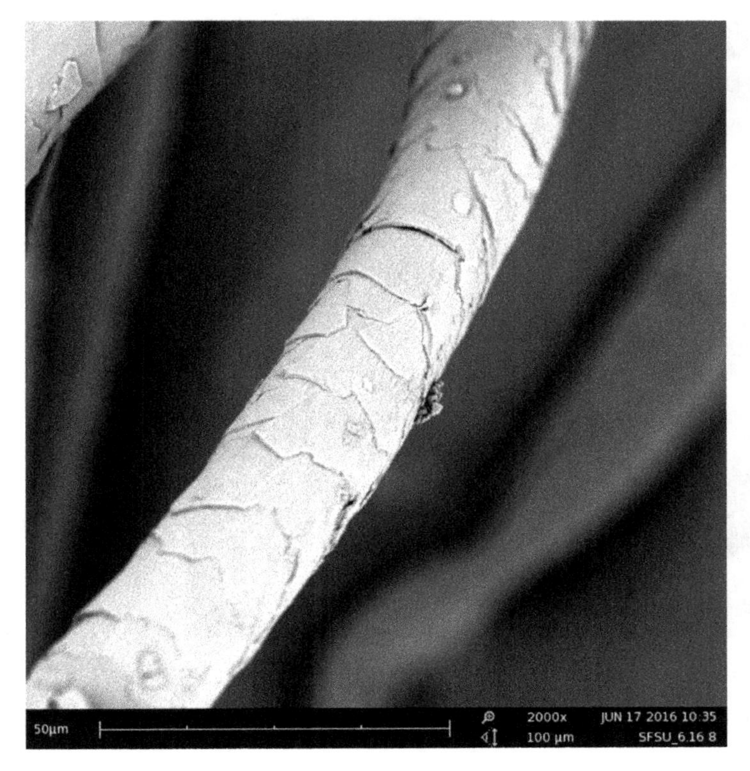

Figure 2.7 Longitudinal view of sheep's wool fiber depicting overlapping cuticle scales (2000×).

height is also larger, around 28 μm, than those of fine fibers. A cross-sectional view of wool fibers shows great variety and helps to understand the shape of the fibers. Coarse and medium wool fibers are usually elliptically shaped, whereas fine wool fibers have a more circular cross-sectional shape (see Figures 2.8 and 2.9). This is true for most wool fiber types. When placing wool fibers on a microscopic slide, it is important that the fibers are slightly straightened because wool fibers have a natural *crimp*. Crimp gives wool fibers wavy and curly shape. The crimp increases the fiber's springiness and bulk, which is why wool fibers are more resilient than other natural fibers such as cotton.

A micron (μ) is the measurement used in describing the diameter (fiber width) of wool fibers. One micron is equal to one millionth of a meter. A lower micron measurement indicates a smaller fiber diameter and therefore a finer fiber. Uniformity in diameter is also very important in fine wool [10]. A comparison of fiber fineness in microns (μ) is listed in Table 2.1.

Figure 2.8 Cross-sectional view of wool fibers showing circular fiber shape (2000×).

Figure 2.9 Wool fibers depicting variety of cross-sections – small circular shapes (fine fibers) and larger elliptically shaped coarse and medium wool fibers. Wool fibers are mixed with U-shaped fibers which are cotton (500×).

2.1.7.1 Fiber Absorbency

Wool has one unusual property for a natural fiber that it is naturally water repellent. Water droplets will be trapped by the scales and stay on top of the fiber and could be easily shaken off. The scales are also covered with natural fat or oils so that the water droplets just slide off of the scales. However, when exposed to water for a longer time period, wool will eventually absorb water in the core of the fiber not on the surface [6].

2.1.7.2 Fiber Shrinkage

Wool's scales (and the scale structure) play a significant role in wool's behavior when it comes to shrinkage and felting of wool fabrics [6]. Scales are arranged in a way that they overlap like scales on a fish, and therefore, they can move only in one direction. When fibers are placed close together and against each

Table 2.1 Comparison of fiber fineness of a variety of animal hairs.

Fiber	Fiber diameter range (μm)	References
Kemp wool	100–250	[11]
Coarse wool	20–70	[6, 10, 11]
Merino wool	10–18	[6, 28]
Cashmere	12–17	[9, 28]
Yangir	13–15	[13]
Mohair	25–40	[2, 28]
Vicuna	13–14	[13]
Camel	17–25	[28]
Alpaca	20–40	[12]
Llama fine	10–35	[12]
Llama coarse	10–80	[12]
Shahtoosh	9–10	[20]
Angora rabbit	11–13	[28]
Yak fine	15–30	[12]

Range in diameter was averaged with two or more sources.

other, they start entangling. This process is called fulling/felting during which fibers will shrink. Wool fibers shrink easily when exposed to water and heat. When wool is wet, the scales will open up and go through a telescoping process (Gail Baugh, personal communication, September 2017. Textile expert San Francisco State University) during which they will shrink.

2.1.7.3 Wool Varieties

Sheep's wool has many variations of wool fibers. The various breeds of sheep produce a variety of wool types. The basic classification of wool is coarse and fine wool. Coarse wool is used more and more frequently in the interior design industry, e.g. in carpets, and fine wool are used in the fashion industry for clothing. Different types of wool will be examined next. For example, *Lamb's wool* is wool that is sheared from sheep aged less than seven months. This wool is softer and finer. *Virgin wool* is defined as wool that has never been processed before. *Recycled wool* (which will be discussed in Chapter 7) is wool that has been shredded into fibers and then used for new textiles. However, the most common type of fine wool is *Merino wool*. It originally came from Spain, but today, this breed of sheep is domesticated in Australia and New Zealand.

2.1.8 Merino Wool and Other Fine Wool Fibers

Merino wool comes from Merino sheep and is the most valuable fine wool. Merino has the softest and finest wool of any sheep, and it also has high luster and good drape.

The scale pattern of very fine wool fibers is the same regardless of breed. Fine Merino fibers, along with other fine wool fibers, will have a scale pattern of the *irregular mosaic* type with smooth margins. Sometimes, the pattern will be of the *waved mosaic variety.* However, in addition to the previously mentioned scale patterns, fine wool fibers such as Merino can also have a coronal or semicoronal pattern in which the scales are large enough to encircle the entire shaft of the fiber. The coronal pattern is not very common in fine wool fibers, and when observed, it is seen closer to the tip of the fiber [2] (see Figures 2.10–2.13).

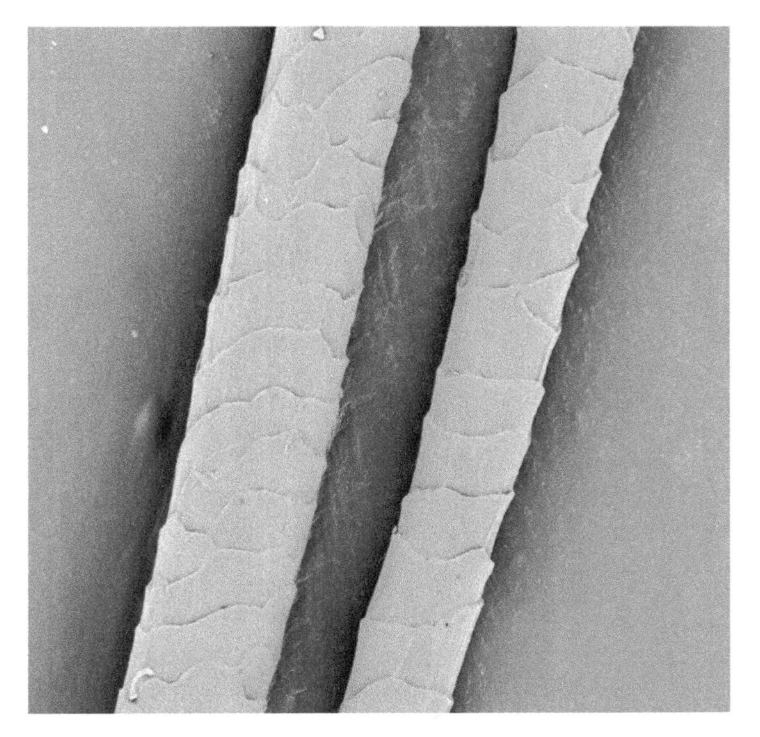

Figure 2.10 Longitudinal view of Merino fibers. Finer fiber (right) depicting semi-coronal to coronal scale pattern where scales encircle entire shaft. Coarser fiber (left) showing larger width and a different scale pattern from the fine fiber (right) (1500×).

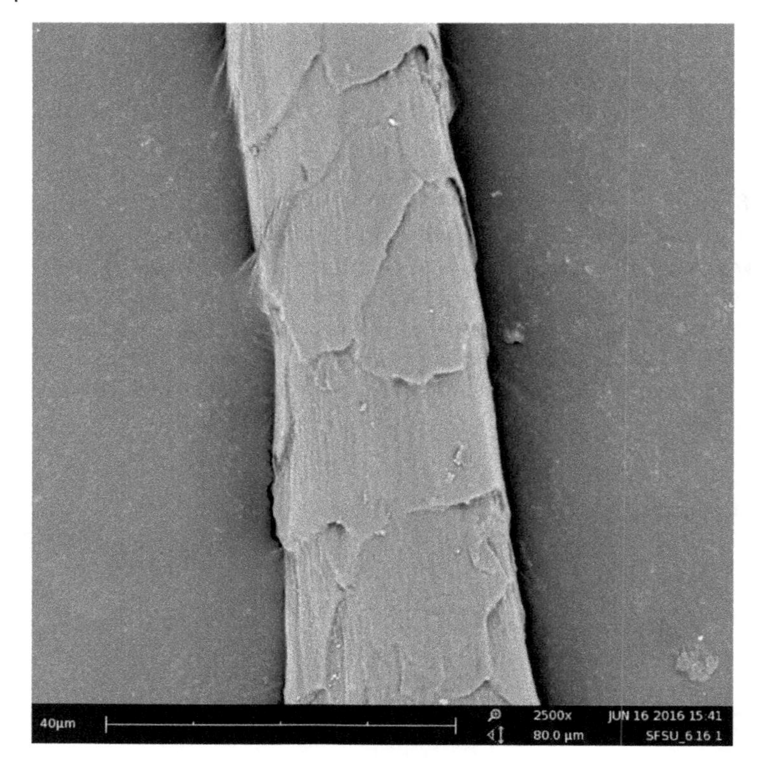

Figure 2.11 Magnified (2500×) view of Merino scale pattern.

2.1.8.1 Normal Fleece Wool

If sheep fibers are medullated, then they are of the wide *lattice* medulla type as in kemp fibers, or of the *simple unbroken* narrow medulla type, which may be seen in some longwool, crossbred wool, or others [2]. The simple unbroken medulla may fade out at points creating an irregularly interrupted medulla or a fragmental medulla. The fragmental medulla is very narrow and does not affect the dying properties of the wool.

2.1.8.2 Kemp Fibers

Kemp fibers are very coarse and therefore undesirable. The characteristics of kemp fibers are a wide medulla, which may be described as a coarse *wide lattice*. Because of the wide medulla, kemp fibers are usually colored white [10]. The medulla, under a transmitted light, is seen as black, and in reflected light, it appears as white [10]. Kemp fibers are believed to be easily distinguishable under a microscope because of their wide medulla and having a knob at the end of the root [2]. Because of the large medulla in kemp fibers, it

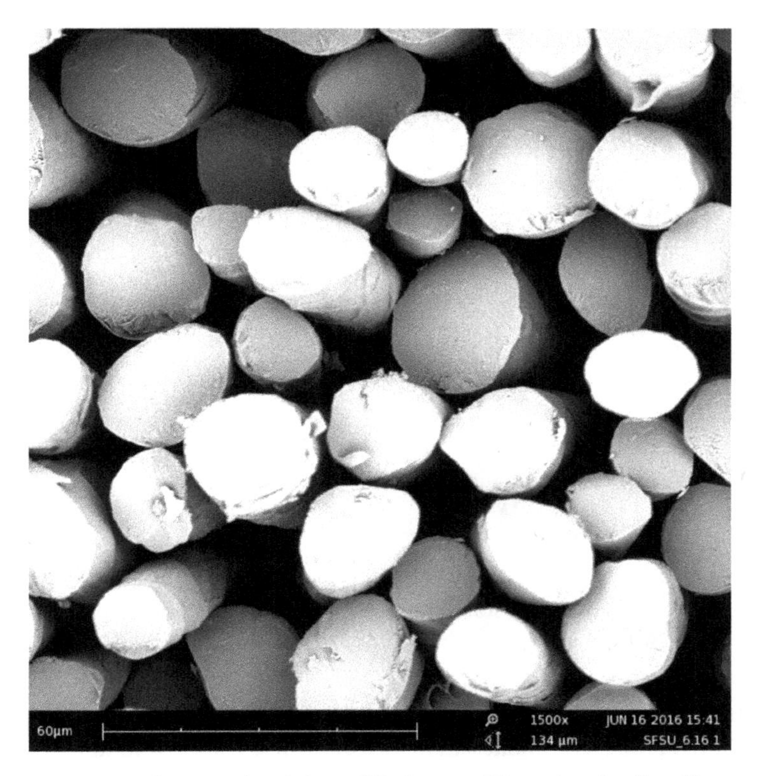

Figure 2.12 Cross-sectional view of Merino wool fibers showing finer fibers and coarser fibers. Finer fibers (smaller in diameter) have a circular cross-section and coarser fibers show circular to oval cross-sectional shape.

does not absorb dyes in the same way the other types of wool do; another reason for being undesirable. Another distinguishing characteristic of kemp fibers is its large diameter as it is larger than other wool fibers approximately 100–250 µm [11].

2.2 Luxury Fibers

A further classification of animal hair fibers is luxury fibers, also called specialty fibers or specialty wool from goats, antelope, yak, ibex, and camelids (to be discussed in the Sections 2.2.1–2.2.9). Luxury fibers, although representing only a small portion of the textile market, are highly sought after and play a very important role in the luxury market [12]. Most of the animals that provide luxury fibers live in harsh mountain terrains and at high altitudes with a harsh,

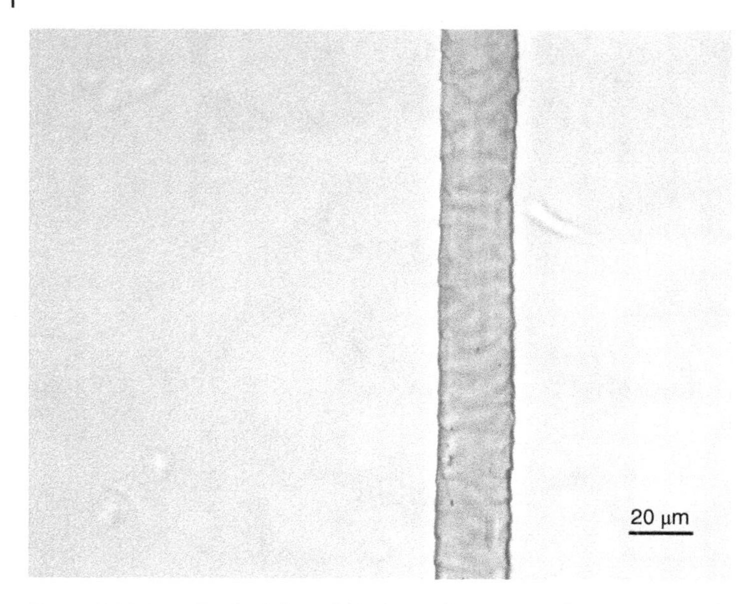

Figure 2.13 Longitudinal view of Merino scale pattern using a compound microscope. Scale shape is not easily identifiable.

cold climate. It appears that because of these inhospitable conditions, animals develop an undercoat of great fineness. Fine luxury fibers are used in apparel for dresses, sweaters, shawls, suits, and coats and in some interior furnishing.

2.2.1 Cashmere

Cashmere fibers (Capra hircus laniger) come from Tibetan goats, whose habitat is in mountainous terrain of Central Asia; in China, Mongolia, Northern India, and Iran. Cashmere is the most luxurious of all wool, and having the smallest fiber diameter, it makes fabric that is soft and smooth. Cashmere's luxurious, soft hand lies in its fibers' fineness. The fineness of fibers is determined by their diameter or width, and cashmere fibers have the lowest diameter compared to all other wool and hair fibers. Although the average fineness ranges from 14 to 17 μm [9], cashmere fibers may have diameters as small as 5 μm. The fiber diameter of coarse fibers ranges from 42 to 160 μm [2]. Whether fine or coarse, cashmere fibers are also very uniform and regular.

As a matter of fact, the cross-section of wool fibers and all other hair fibers is circular in shape with a visible epidermis, cortex, and medulla. The cross-section of fine wool reveals a more uniform circular shape of fibers, whereas the circular shape of coarse wool fibers is not as uniform and not always circular, i.e. circular to oval. A medulla is not present in fine fibers. When distinguishing fine wool, cashmere, and mohair from coarse wool, the fiber diameter or width is an important consideration. The finer the fiber, the smaller the diameter of the fiber, and the coarser the fiber, the larger the diameter of the fiber.

Because of the high price of cashmere wool, it is sometimes blended with fine wool to reduce the cost. Therefore, it becomes very important to able to distinguish the two, fine wool and cashmere wool. Specialists focus on scale formation when distinguishing wool fibers.

Cashmere fibers may also be fine and coarse. Fine cashmere fibers are not medullated and have cuticle scales that are not prominent [2]. The scale pattern type is waved mosaic, and the wave is not always regular as it may be interrupted [2]. One distinguishing characteristic of fine cashmere fibers is that the scale margins are distant and scales are far apart, which makes the fibers smooth [2, 6] (see Figure 2.14). Coarse fibers, on the other hand, have an unbroken type of medulla present, and the scale pattern is of an irregular-waved mosaic pattern with near and crenate-rippled margins [2]. There are

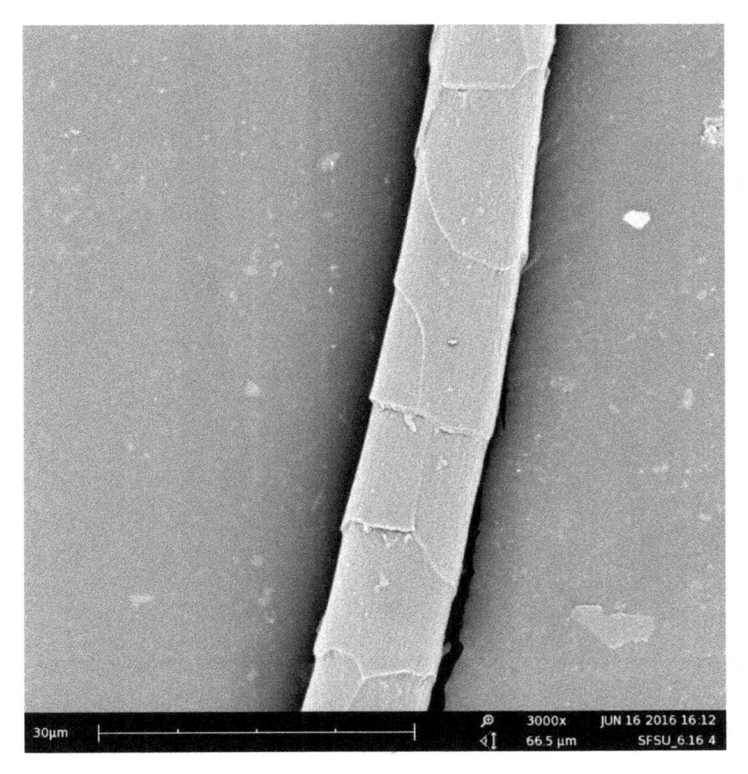

Figure 2.14 Longitudinal view of cashmere fine fiber depicting far apart scales which make the fiber smooth (3000×).

also fibers of intermediate thickness, which also probably do not have a medulla. These intermediate fibers are more similar to fine cashmere fibers than to coarse cashmere fibers especially because they have no crenations present. They have smoother margins just like fine fibers but have an irregular wave mosaic pattern just like coarse fibers [2] (see Figure 2.15).

All of these types of cashmere fibers are regular in diameter. The cross-section of cashmere fibers is circular for fine fibers and of oval-shaped for coarse fibers (see Figure 2.16).

Cashmere fibers have special pigment granules unevenly distributed in its cortex, which can be clearly seen when viewing the fiber cross-section using the light microscope.[1] These pigmented granules are what distinguish cashmere fibers from merino and mohair fibers, because the latter two do not have pigment granules present.

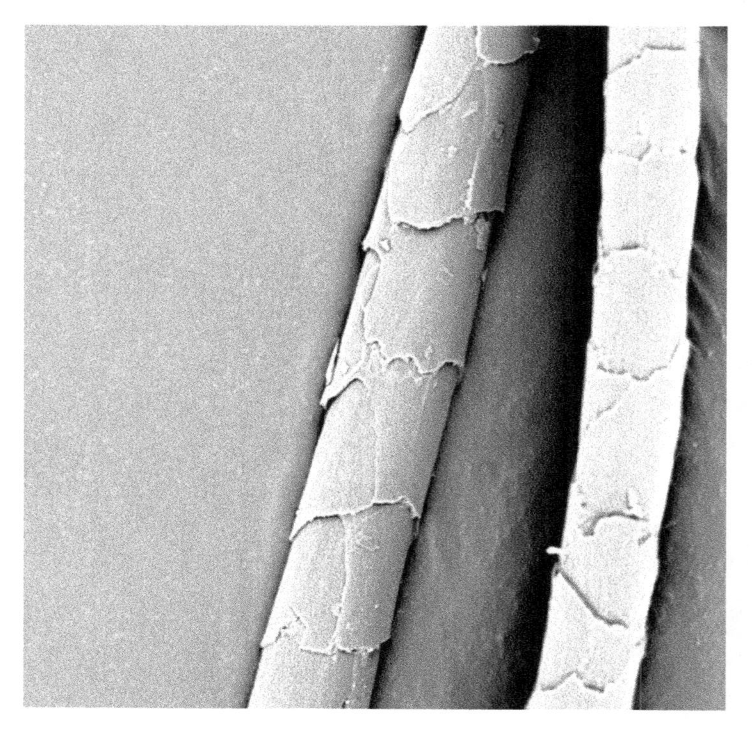

Figure 2.15 Cashmere fibers of intermediate thickness (left) have an irregular wave mosaic pattern just like coarse fibers (2500×).

1 Only SEM images of cashmere are provided in this textbook in which these granules cannot be seen.

Figure 2.16 Cross-sectional view of cashmere fibers depicting a variety of cross-sections stemming from fine (circular shape) to coarser fibers (shape not uniform) (2500×).

2.2.2 Yangir

Yangir, also called the wild cashmere, mainly comes from an Asiatic subspecies of Capra ibex, the wild goat Siberian ibex C. ibex sibirica. The wild goat lives in mountainous areas from India to Mongolia. Although it is named after Siberia, it does not live there. Just as with the Shahtoosh, the animals must be killed to obtain the fiber. Dehaired yangir fibers are fine and short with a diameter around 13–15 µm. The fiber's natural color is light blonde [13]. The scales of yangir fibers appear to be elongated in the direction of the fiber axis, somewhat similar to Shahtoosh fibers. The distinguishing characteristic of yangir fibers is the thickness of the scale edge, which is thicker than that of cashmere fibers. There are around 5–6 to 18–20 scales in 100-µm fiber length [14].

2.2.3 Mohair

Mohair (C. hircus aegagrus) is another specialty fiber that comes from the Angora goat found in Tibet. Mohair fibers are circular and uniform with scales

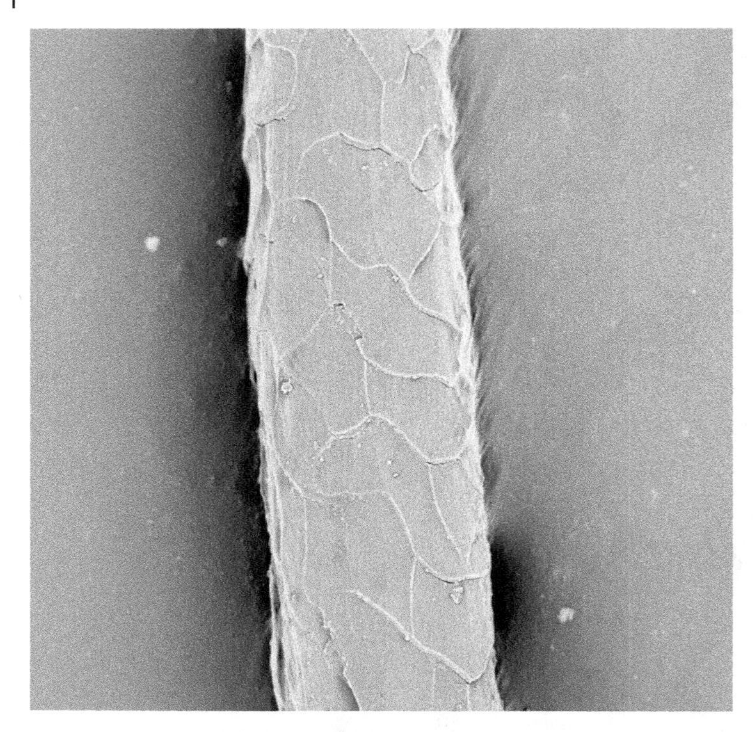

Figure 2.17 Longitudinal view of Mohair fiber showing scales being tight in relation to one another, giving mohair fiber smoothness, luster, and low friction. Scales are less pronounced not as visible (1500×).

that appear to have a smooth surface [15] and to be flatter on the fiber (Figure 2.17). Mohair fibers are very even in diameter with scales that are difficult to see because they are thin [2] (see Figure 2.18). When mohair fibers are compared to fine wool fibers of the same diameter, the mohair fibers are straighter and smoother than wool fibers [10]. The microscopic structure of mohair is similar to wool [7] as the mohair scale pattern is more similar to wool's scale pattern, differing from the comparison of cashmere fibers to wool fibers. When compared to wool, scales on mohair are less pronounced and not as visible [16]. Mohair fibers do not have as many scales as wool fibers [1], and therefore do not shrink nor felt as freely as wool fibers [6]. The scales are tight in relation to one another, giving mohair fiber smoothness, luster, and low friction. The scales of mohair are thinner (around 0.4 μm) than those of wool, and there are fewer of them in mohair than they are in wool. There are around 5–7 scales in 100 μm of fiber in mohair compared to 9 scales in fine wool. The scales of mohair also do not overlap as much as those of wool. Because of this

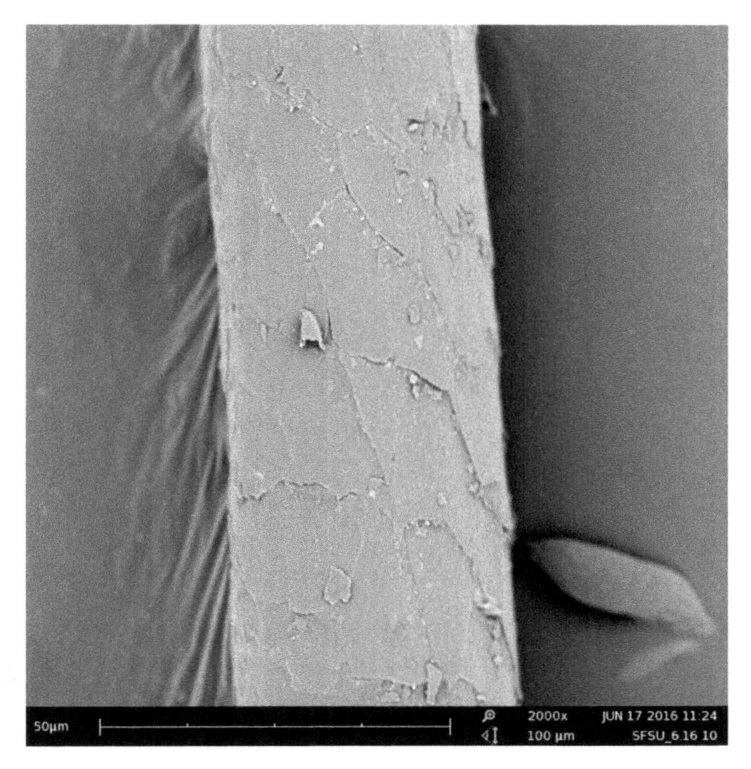

Figure 2.18 Longitudinal view of brush Italian mohair fibers depicting less-pronounced scales positioned close to the fiber shaft (2000×).

low-scale distribution, mohair fibers are more lustrous and smoother than wool fibers. Also, because of its luster, mohair makes good wigs [1].

Only small percent of mohair fibers are medullated [7]. The medulla is interrupted or fragmented [17] (see Figures 2.19 and 2.20), and in some samples, fibers exhibit an opaque medulla as in kemp fibers [10]. In mohair fibers, vacuoles running longitudinally to the cortex have also been identified [17]. When scientists want to distinguish between specialty fibers and wool fibers, they also consider the size of the scales as it is believed that the scale size is larger in specialty fibers such as mohair [17] than in wool. Fibers with a scale size below 18 µm would be classified as wool, and fibers with a scale size above 18 µm fibers would be classified as mohairs [18].

The fiber diameter of mohair fibers varies from 25 to 26 µm in fine fibers and up to 55 µm in coarse fibers. Mohair fibers also have both fine and kemp fibers. For the most part, mohair fibers are not medullated except for the kemp fibers

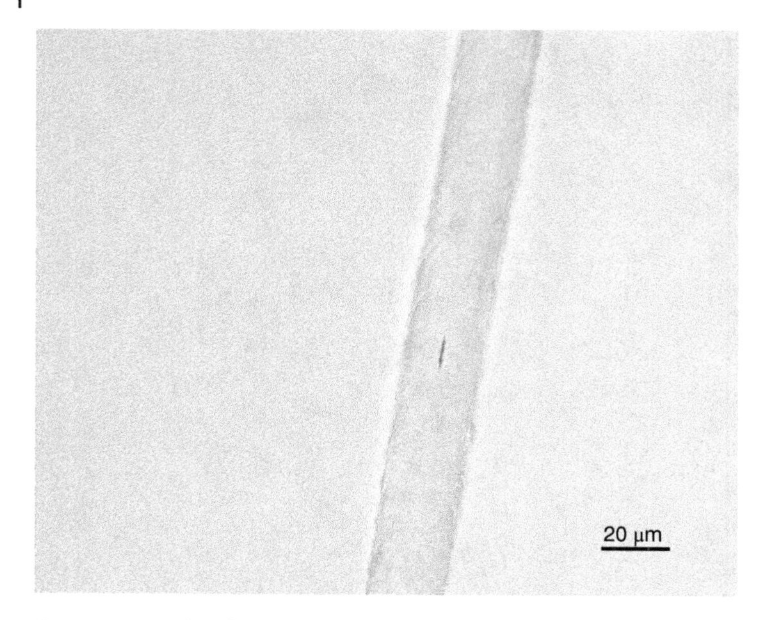

Figure 2.19 Mohair fiber (compound microscope) depicting a fragmented type medulla (notice short black line in the middle of fiber).

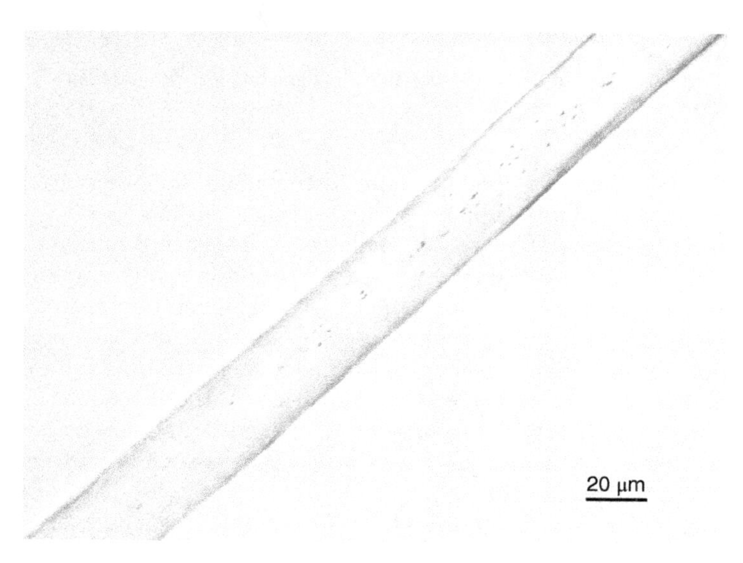

Figure 2.20 Mohair fiber showing fragmented medulla via compound microscope. Medullae are not present in fine fibers; they are only present in coarse or kemp fibers. Only small percent of mohair fibers are medullated.

that have a wide lattice type medulla. The sequence of the scale pattern of mohair kemp fibers is not of the same type as wool kemp fibers [2]. Coarse mohair fibers are also medullated, having an interrupted medulla.

The cross-section of mohair fibers is circular for fine fibers and elliptical or ovoid for larger, coarse fibers (see Figure 2.21). Mohair's cross-sectional view has no unique or distinguishing characteristics that would help in identification [2]. The cross-section of mohair fibers is very circular with black dots that are air-filled pockets [7]. These dots are not the pigmented granules we see in a cashmere cross-section.

The cuticle scale pattern of mohair fibers is that of the irregular waved mosaic with the waves being somewhat interrupted. The mosaic pattern varies, being sometimes narrow and elongated lengthwise [2]. When it comes to kemp mohair fibers, two different scale patterns can be observed: the simple wave pattern and the irregular slightly waved mosaic pattern with smooth and near margins. This pattern may also be observed on the tip of sheep's wool

Figure 2.21 Cross-sectional view of brush Italian mohair fibers depicting fine and course fibers. One medullated fiber (left bottom) depicting cellular nature of medulla (2000×).

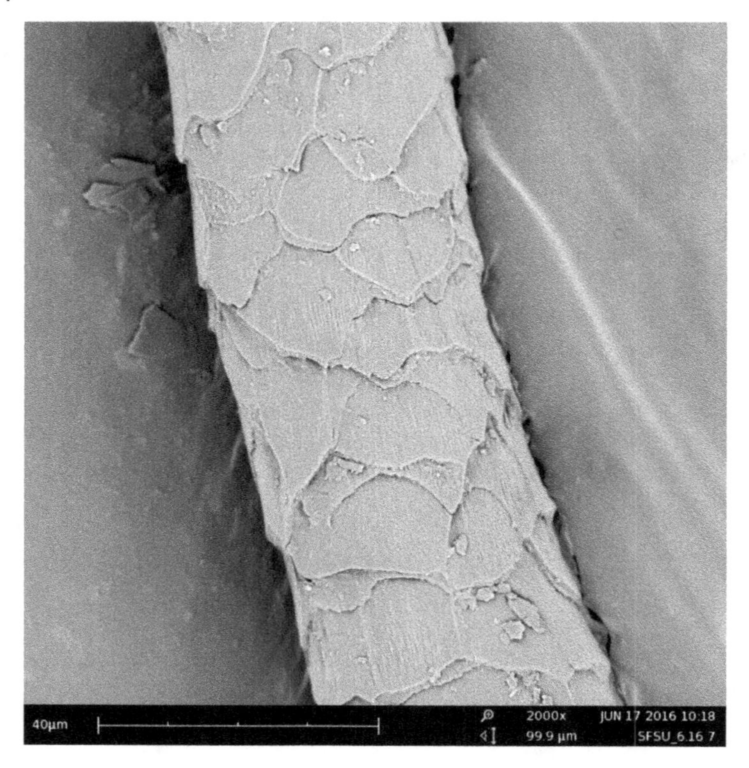

Figure 2.22 Longitudinal view of Mohair intermediate (to coarse) fibers, and the scale pattern is of the irregular mosaic (2000×).

kemp fibers. When it comes to coarse and intermediate fibers, the scale pattern is of the irregular mosaic [2] (see Figure 2.22). Mohair is used for suitings, sweaters, and interiors such as throws, upholstery, and draperies (Figure 2.23).

2.2.4 Vicuna

While alpaca and llama are domesticated camelids, vicuna and guanaco live in the wilds of Bolivia and Peru and are protected today by the Endangered Species Act of 1999. Therefore, apparel items containing vicuna and guanaco fibers are not legally allowed to be brought into the United States. While vicuna fibers are similar to those of alpaca and llama, vicuna fibers are finer [10]. Vicuna is the finest, softest, and also the rarest fibers in the world, and therefore, they are the most expensive [1]. The average diameter of vicuna fibers is approximately 13–14 μm, and the scales appear to have a smooth-margined irregular waved pattern; the scales' pattern is coronal and distant [13]. Vicuna fibers do not have as many scales (approximately 6–7 scales per

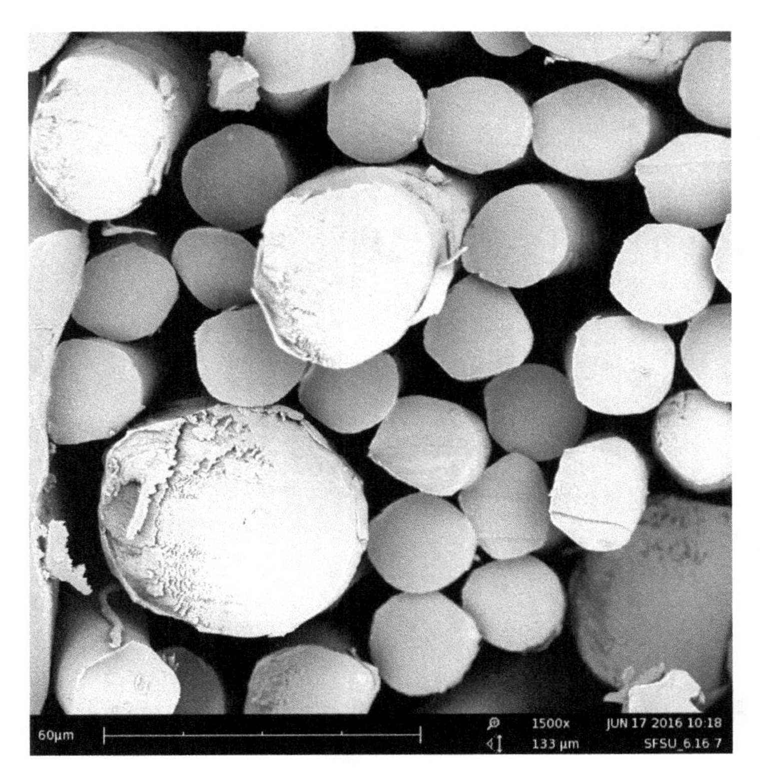

Figure 2.23 Mohair fibers depicting fibers of various diameter with circular cross-section and no medulla present (1500×).

100 µm) as those of alpaca. There is a medulla in vicuna fibers just as in all other camelid fibers. However, it is sometimes absent in the finest fibers. The medulla is visible in coarse and fairly fine fibers. It is opaque [10], fragmented, or interrupted [13].

2.2.5 Camelid Fibers

Camelid fibers are a type of hair fibers that are also considered to be luxury fibers. Under the microscope, they have the characteristics of hair. Unlike cashmere and mohair, most camelid fibers are medullated. The hollow shaft, medulla, might have a honeycomb structure that is continuous or fragmented/interrupted, or be completely empty (see Figure 2.25).The hollowness adds the unique qualities of lightweightedness, warmth, and coziness to the end product. Not all camelid fibers are medullated though (see Figures 2.24 and 2.25). Some fibers have the characteristics of wool and some of hair [7].

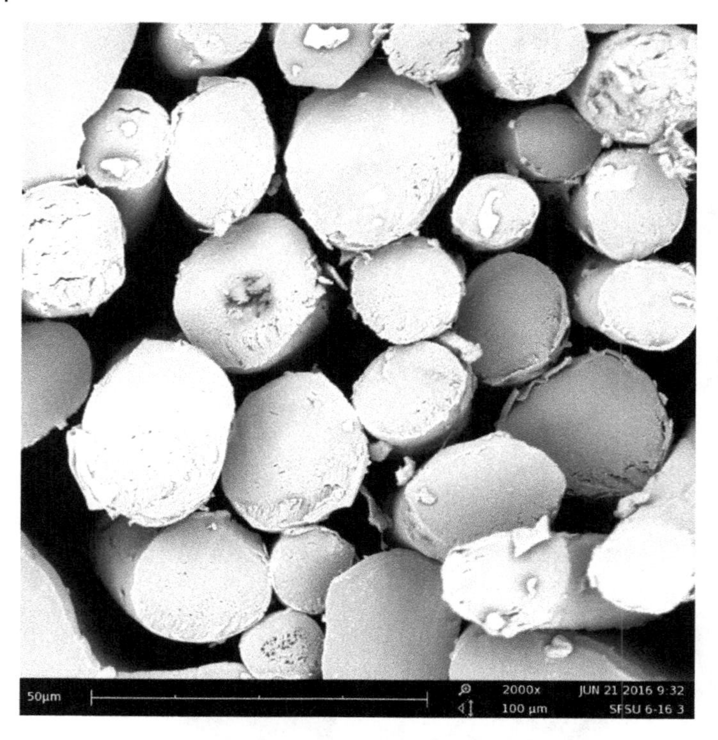

Figure 2.24 Cross-sectional view of camelid fibers depicting unmedullated and medullated fibers in varying degrees of finesse (2000×).

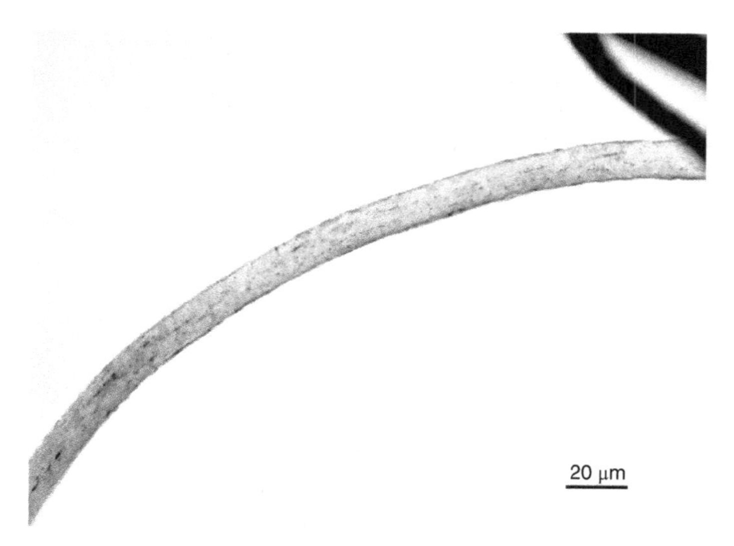

Figure 2.25 Medullated camel fiber showing fragmented medulla via compound microscope.

Figure 2.26 Camelid fibers contain less scales and scales are far apart making people allergic to wool not always be allergic to camelid fibers (2000×).

Camelid fibers do not contain as many scales as wool does, and this makes the fibers feel softer. This is also one of the reasons why people allergic to wool are not necessarily allergic to camelid fibers (see Figure 2.26). The scales of camelid fibers are long and not visible (see Figure 2.27). There are about 7–9 scales in 100 μm [7]. It is interesting to observe that the edges of the scales are sharply bent.

2.2.6 Alpaca

Alpaca fibers come from a camelid found in the Andean Regions. Alpaca fibers are long, soft, and have luster. Alpaca fibers are used in their various natural colors; gray, fawn, white, brown, and black [13]. The fiber diameter is between 20 and 40 μm (Figure 2.28).

The scale pattern differs based on the fineness of the fibers. The scale pattern of alpaca fine fibers is semicoronal in shape, which seems as if it is part of an irregular-waved mosaic pattern [2] (see Figure 2.29). The scale pattern is

Figure 2.27 Longitudinal view of camel fiber showing long scale structure (2480×).

believed to be similar to that of fine wool's pattern, but they differ in the external margin scale pattern. The scales do not project away from the fiber in alpaca fibers as they do in wool fibers and therefore appear smoother [2]. The coarse fibers have an interrupted regular wave scale pattern with scale margins not smooth but crenate-rippled. Intermediate thickness fibers have an irregular wave pattern and irregular wave mosaic scale patterns [2].

The appearance of alpaca fibers and wool fibers may seem similar, which may make the task of differentiating or identifying them very difficult. The scales of alpaca, both Huacaya and Suri types, are different from wool scales even though they look very similar to wool under the microscope. For example, alpaca scales are thinner and denser when compared to wool fibers [19]. When analyzing both scale height and number of scales per 100 μm, differences were observed. Alpacas have a lower scale height (approximately 0.50 μm) than that of wool (0.65 μm), as is expected in finer fibers. However, when analyzing the number of scales per 100 μm, alpaca fibers have more scales (approximately 8–10) than sheep's wool fibers (approximately 7). It is usually the opposite when comparing fine fibers to wool fibers.

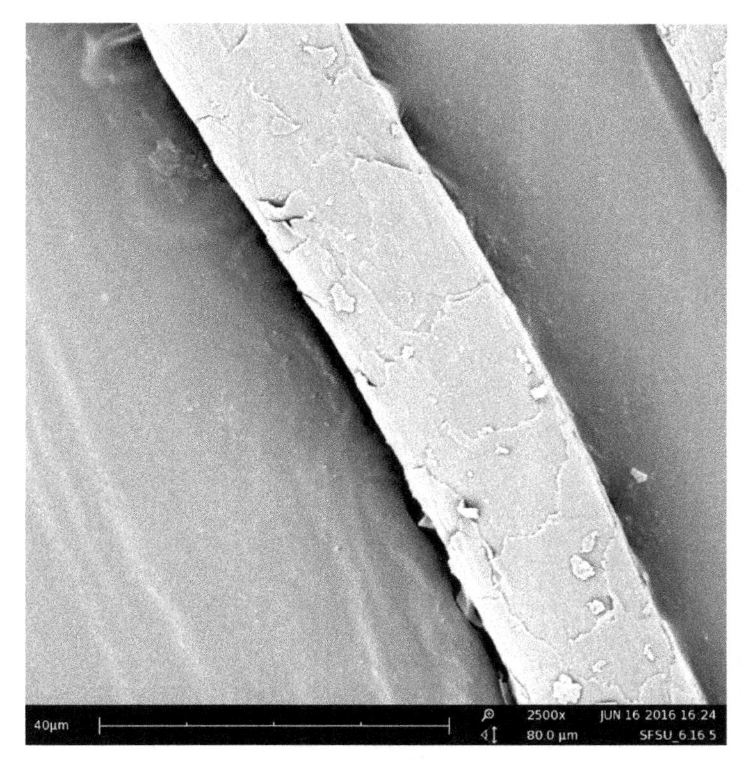

Figure 2.28 Magnified (2500×) longitudinal view of alpaca fine fiber depicting semi-coronal scale formation. Alpaca fibers often contain somewhat indistinct scale formation.

The fleeces of camelids, e.g. alpaca and llama, are similar [10], and it becomes somewhat more difficult to distinguish the two under the microscope. Both of these fibers are also similar to mohair from Angora goats [7], more specifically in scale formation.

A simple, narrow, unbroken medulla is usually present in alpaca fibers. Sometimes, the medulla is unbroken for most of the length and then changes to an interrupted or a fragmental medulla [2]. Some experts say the most important characteristic of alpaca and llama fibers is that many have an interrupted medulla present [7] (Figure 2.30).

Llama fibers are fine and smooth, and the scales are scarcely distinguishable [7, 10]. Usually, such fine hair fibers would not have a medulla present, but llama fibers are characterized by a narrow, opaque medulla.

There are pigment granules in the cortex of colored fibers, which give the appearance of dark spots in the cross-section. These are present regardless of the fiber color.

Figure 2.29 Longitudinal view of alpaca fibers (1500×). The scale pattern of alpaca fine fibers is semicoronal in shape that seems as if it is part of an irregular-waved mosaic pattern.

Figure 2.30 Alpaca fiber depicting interrupted medulla via compound microscopy.

Figure 2.31 Alpaca fibers have differing cross-sectional characteristics depending on fiber fineness – the coarser the fiber the more elongated the fiber cross-section creating dumb-bell shaped medulla (1500×).

Alpaca fibers of different colors may also have differing cross-sectional characteristics. Also, the coarser the fibers get, the more the shape of medulla changes. For example, in medium to coarse fibers, the medulla is elongated. In coarse fibers, the medulla is dumb-bell shaped, and in very coarse fibers, the medulla can be of T-shape or other unusual shapes [2] (see Figure 2.31). These unusual appearing medulla shapes, seen in cross-section, are reliable indicators that these are camelid fibers from camels such as alpaca or llama.

2.2.7 Llama

The llama (Lama glama glama) is a camelid living in the high altitudes of the Andes Mountains. Llama fibers are of both finer and coarser types. The finer types range in diameter 10–35 µm. It is interesting to note that only few of the finer fibers are not medullated. Many of the fine llama fibers have a narrow medulla of the interrupted type with scarce interruptions. When analyzing the differences between llama and alpaca fibers, there are some things to note [2].

For example, llama fibers seem coarser than those of alpaca, and therefore, the llama fleece is not as attractive as alpaca fleece [12]. The cuticle scales of fine llama fibers are of the irregular waved mosaic type with smooth and near margins. The cuticle scale pattern of coarse llama fibers is also of irregular-waved mosaic pattern of the interrupted type. The margins are not as smooth as those seen with smooth fibers but are margins of a rippled-crenate type with scales that are close to each other [2]. The cross-section of llama fibers is somewhat similar to the cross-section of coarse alpaca fibers in that the medulla shapes are unusually formed, and some fibers have an elliptical cross-section. However, when a bunch of llama and alpaca fibers are compared in their cross-section, they can be distinguished because llama fibers have more unusually shaped medullae. Llama fibers have medullae that are tripartite and quadripartite.

2.2.8 Shahtoosh

Shahtoosh is another fiber that comes from the hair of a wild antelope (Pantholops hodgsonii) living in Tibet. Shahtoosh fibers are considered to be premium and the finest fibers in the world. However, a problem is that in order to obtain the fibers, the antelopes must be killed [13]. The fine Shahtoosh fibers, which come from the undercoat of the wild animals, have a fiber diameter less than 10 μm [20]. As other fine animal fibers, they do not contain a medulla. The scale pattern of fine Shahtoosh fibers is coronal and varies in distance along the shaft. The scale features of fine Shahtoosh fibers are very unique as the scales taper to a point. This is very different from cashmere scale ends because they are blocky in appearance and closer together. The scales of Shahtoosh fibers are quite distinguishable as they are elongated in the direction of the fiber axis, and their lengths, as well as their thickness at the distal edge, are pronounced and relative to the fiber diameter [21]. There are even fewer scales per 100 μm in Shahtoosh than in previously mentioned hair fibers. Shahtoosh has only 5–6 scales per 100 μm [22]. Because of these fiber characteristics, the fiber diameter may not be as regular along the fiber axis [21].

Coarse Shahtoosh fibers are easily distinguishable because they have two distinctive characteristics, a continuous medulla that completely fills the fiber shaft and a large rounded shape of the large medullar cells [20]. There is quite a difference in Shahtoosh's coarse fibers' cuticle scales and fine fibers' scale formation.

2.2.9 Yak

The yak (Bos grunniens) is a member of the hoofed Bovidae family, and its habitat is in high altitudes of the Tibetian' Plateau [12]. Yak's fine down hair is very fine and can be easily mistaken for that of cashmere, and therefore, it is of interest to be able to distinguish these two. It is interesting to note that yak hair is half the price of cashmere's. The diameter of yak hair is around 15–30 μm for fine (down) hair and around 35–80 μm for coarse hairs [12]. Yak fibers have a larger diameter than those of cashmere fibers. Under a microscope, the fibers appear to have oval to circular cross-sectional shape (see Figure 2.32) and have an interrupted medulla (Figure 2.33) or no medulla. Scale formation is very similar to those of cashmere fibers. A good distinguishing factor is that yak fibers have a higher scale frequency (9–10 scales) when compared with cashmere (6–8 scales) fibers [5] (see Figures 2.34 and 2.35 for comparison). Yak fibers are used for warm clothing such as sweaters or panchos.

2.2.10 Other Identification Techniques to Note

When identifying animal fibers from each other or within a group of animal fibers, some researchers suggested a multiparameter approach that incorporates not only identifying the overall scale pattern of fibers and diameter but also the *scale height* and *scale frequency* [5]. The demand and

Figure 2.32 Cross-sectional view of Yak fibers depicting oval (coarser fibers) to circular (fine fibers) fiber shape (2000×).

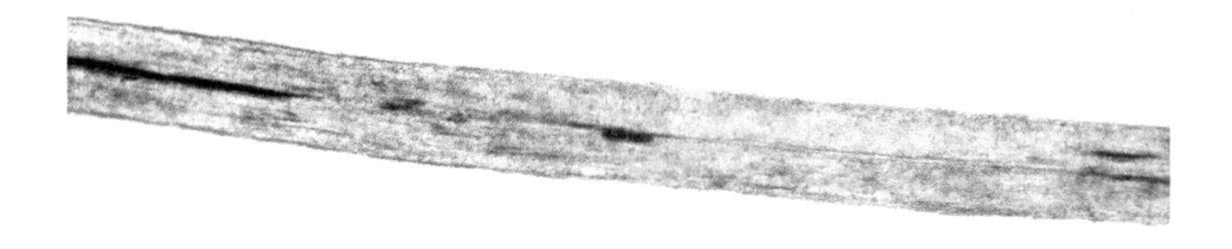

Figure 2.33 Longitudinal view of Yak fibers depicting an interrupted medulla.

price for luxurious specialty fibers such as cashmere drive the fraudulent blending of these high-priced specialty fibers with lower-priced wool. Therefore, it becomes important to differentiate specialty fibers, such as cashmere, from wool. Their scale pattern differs as cashmere's scale pattern is described as a scale that is thin [9]. In cashmere, the scale edge is smooth and clear, and ring-like to cover the hair shaft, while in fine wool the scale is thick, with a coarse surface. In cashmere fibers, some scale edges are ring-like and the edge of scale is wide and dim.

Besides differences in scale pattern, the cuticle *scale height* also differs in wool fibers and cashmere fibers when viewed under a SEM. Scale height is categorized as low and high and is based on a 0.6 μm midpoint [23, 24]. Specialty fibers would have a scale height below 0.6 μm, while wool fibers would have a scale height above 0.6 μm. The mean diameter of fibers is yet another parameter that distinguishes specialty fibers and wool.

Scale frequency is another parameter to be taken into consideration. Scale frequency is defined as the number of scales on a fiber length of 100 μm [5]. What this means is that when viewing a fiber under a microscope, the scales might be more spread out or more close together. If the scales are closer together, it means there are more of them within a certain length and when they are more spread out, there are less of them within a certain length. This scale frequency parameter is also used to differentiate luxury fibers. Table 2.2 illustrates a variety of luxury fibers and their corresponding scale frequency values. For example cashmere fibers and yak fibers have a similar scale formation and thus difficult to distinguish under a microscope, however, cashmere fibers have fewer scales than yak fibers (see Figures 2.34 and 2.35 for comparison).

Table 2.2 Luxury fibers scale frequency values.

Fiber type	Mean scale frequency (100 μm)
Cashmere	6–8
Mohair	6–7
Camel hair	6–8
Alpaca	10
Llama	10
Yak hair	9–10

Source: From Phan et al. 1987 [5].

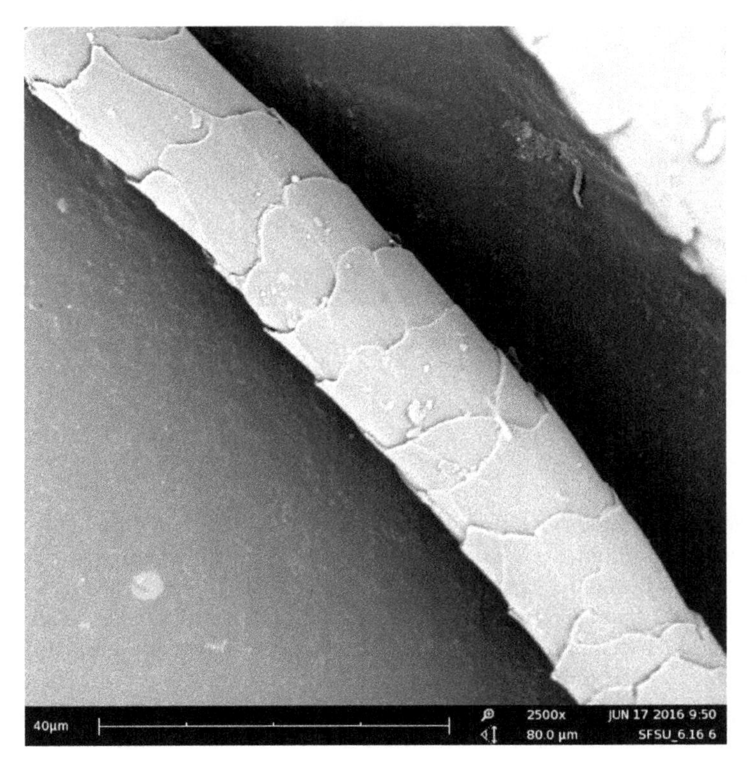

Figure 2.34 Longitudinal view of Yak fibers depicting (2500×) a high scale frequency when compared to cashmere fibers.

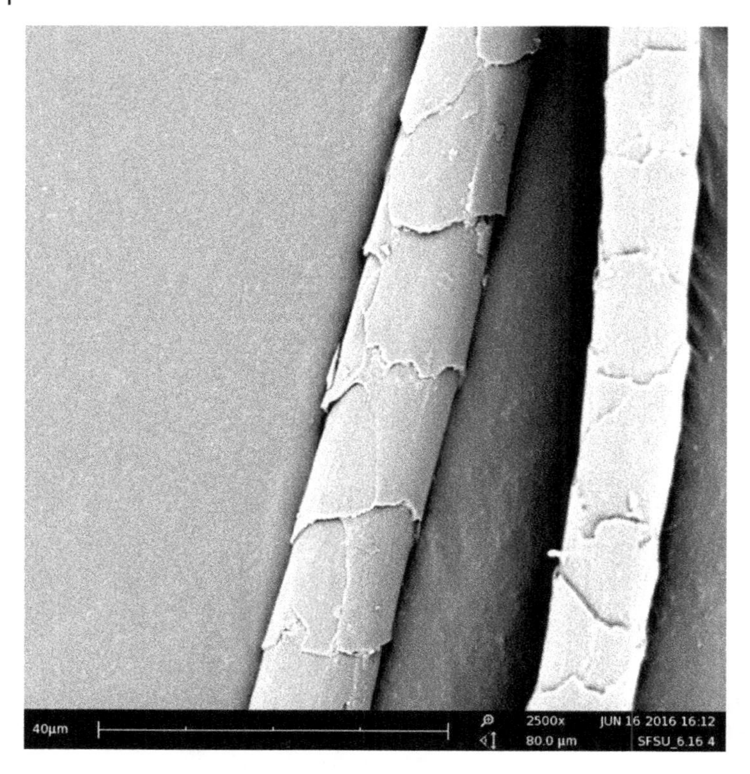

Figure 2.35 Longitudinal view of cashmere fibers depicting cashmere's scales as being further apart than Yak's scales (2500×).

2.3 Silk

Silk is an ancient fiber the usage of which dates back to 4000 CE in China. Silk is the only natural filament fiber used that is secreted by silkworm *Bombyx mori*. The silkworm builds its cocoon by extrusion of the filament in a long continuous strand. When this process is controlled and domesticated through a regiment of feeding the silkworm only mulberry leaves, it is called sericulture. The cocoon has to be unwound to get the long filament fiber. To get an unbroken long filament strand, the cocoons are boiled (larvae is killed to keep it from breaking through the cocoon) and then the cocoon is unwound. There are two types of silks: cultivated and wild. Cultivated silk (*B. mori*) is the more commonly used silk because it is finer and considered of higher quality. Wild silk, on the other hand, is not domesticated and thus is not as uniform as cultivated silk. These two types of silk varieties have distinguishable characteristics under the microscope.

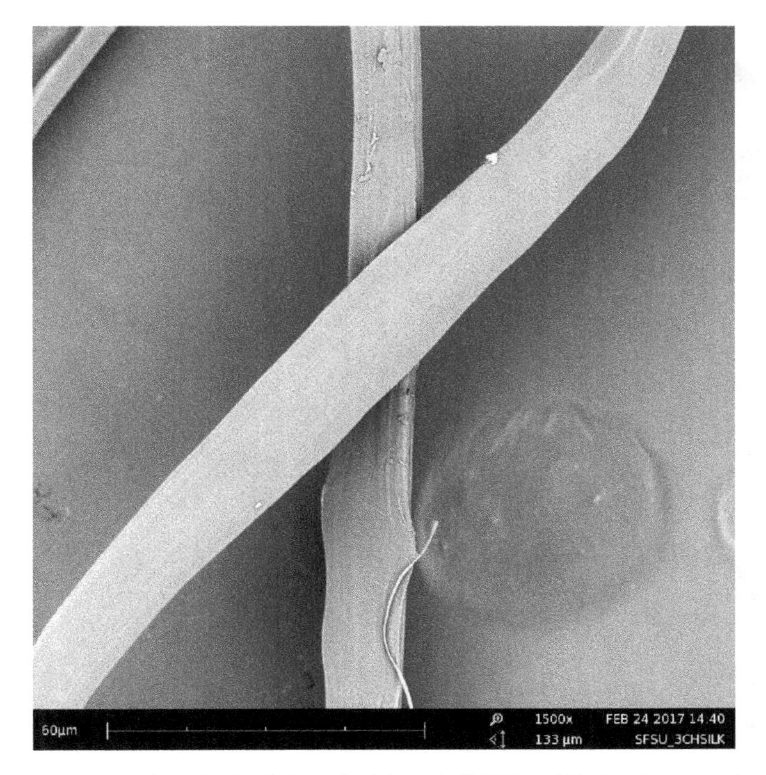

Figure 2.36 Longitudinal view of cultivated silk (China silk) showing smooth surface (1500×).

A) Cultivated silk has a smooth longitudinal surface, a smooth rod, and has a triangular cross-section (see Figures 2.36 and 2.37). This triangular cross-section becomes very important because it plays a role in the luster of silk fabrics. Luster derived from the way light is reflected from the surface of the fiber. The distinct edges of the more triangular shape of the cultivated silk fiber allow the light to evenly reflect. Cultivated silk used to be easily distinguished from all other fibers because of its narrow diameter, but with the development of microfibers, it is becoming more difficult as its smooth rod could be easily mistaken with nylon [25].
B) Wild silk – the two wild silk varieties are the Tussah and the Anaphe. Tussah is the most commonly used of the wild silks. Wild silk fibers can be distinguished from cultivated silk fibers under the microscope, especially when examining the cross-sectional shape. Tussah silk's longitudinal shape is broader with lengthwise striations (the illustration in Figure 2.38) has a ribbon-like, flat cross-sectional appearance [25]. The second type of wild silks,

Figure 2.37 Cross-sectional shape of cultivated silk (China silk) depicting triangular cross-section. Triangular cross-section becomes very important in cultivated silk – it adds luster to silk fabrics (1500×).

Anaphe silk, can even be distinguished from Tussah silk because it often has cross-striations at intervals along the fiber [26]. The cross-section of Tussah silk is wedge-shaped and not triangular as seen in cultivated silk (see Figure 2.39). Moreover, the cross-section of Anaphe silk is roughly triangular, but the triangular shape is slightly different from the cultivated triangualr shape. In Anaphe silk, the apex of the triangle is elongated and bent. Therefore, Anaphe silk ends up being crescent-shaped in cross-section, "the crescent being formed by two curved triangles, joined along their base" [26, p. 564].

Silk is a protein fiber that is composed of two types of proteins: fibroin and sericin. Fibroin is the fiber itself and sericin is the substance that surrounds the fiber. Since silk fibers come in two strands, the sericin substance holds the two strands together. Normally, the two strands are separated by the process called degumming. Degumming will wash the sericin off, so all that is left is the

Figure 2.38 Longitudinal view of wild silk showing some striations (2000×).

fibroin or the fiber. Degumming will improve silk fibers and fabrics will have more sheen. The silk fibers that have not been degummed are called raw silk. The cross-section of raw silk has an elliptical shape because these are the two fibers (two strands) with the sericin holding the triangular twin fibroin or strands together (see Figure 2.40). The longitudinal shape of raw silk is also striated. Some researchers claim that striations could be sometimes caused by degumming of silk fibers. "After being degummed, the individual filaments showed a surface texture consistent with an oriented fibrillar structure in the fiber interior" [27, p. 8860].

2.3.1 Peace or Ahimsa Silk

Silk cultivation (sericulture) is considered an inhumane practice by many today and therefore a new type of silk was created called *peace* silk or *Ahimsa* silk. The production of peace silk does not require killing of the larvae in the process. In this process, the cocoon is allowed to mature into a chrysalis and the

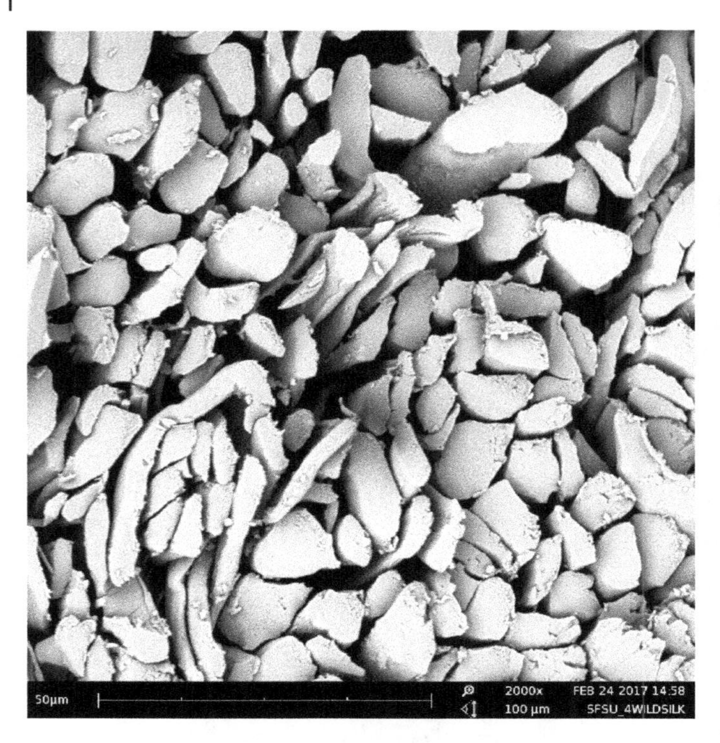

Figure 2.39 Cross-sectional view of wild silk showing wedge-shaped not triangular fiber shape as seen in cultivated silk (2000×).

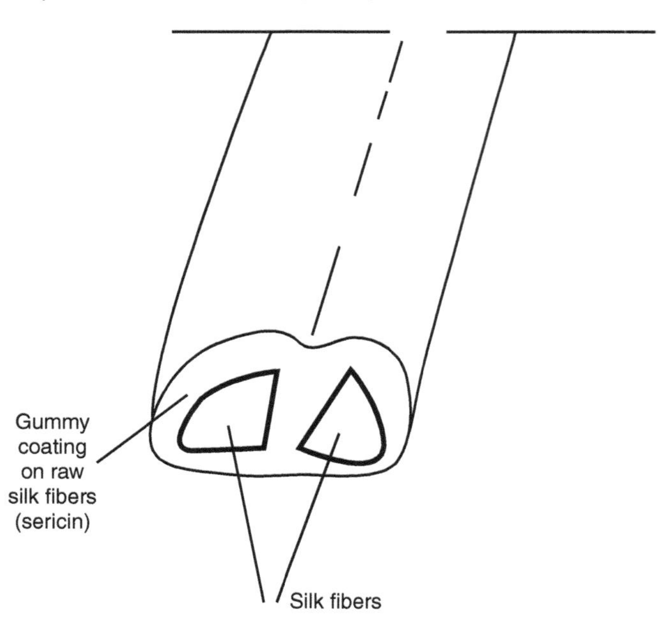

Figure 2.40 The cross-section of silk depicting elliptical shape – two fibers (two strands) with sericin holding the triangular twin fibroin or strands together.

moths are allowed to escape from the cocoon prior to the silk production. As a result we end up with a punctured cocoon that yields broken, short silk fibers as opposed to long continuous silk filaments. Because of its short fibers, peace silk has lower sheen and luster and would look more similar to raw silk fibers under a microscope. The production process is more labor-intensive; therefore, the fabric would be more expensive than regular silk. The fibers are cream to brown in color and the fabric is soft, but might contain bumps.

2.3.2 Spider Silk

Spider silk is another protein fiber that comes from animal secretions. Spiders create silk to make spider webs, which have very desirable properties. Because the purpose of the web fibers is to trap insects, the most important property is its strength. Spider fibers have great tensile strength, five times that of steel and could be used in protection wear such as bullet-proof vests. Additional desirable properties are its extreme lightweight and elasticity, both are important properties in wearables. Silk fibers absorb energy without breaking; when a fly flies directly into a web, the web does not break. Without the spider fibers' ability to absorb energy, the web would break. This is, for example a valuable property for use in cycling helmets. Another property associated with spider silk fibers is super contraction, when wet the fibers shrink to almost half its size. Because of all of these outstanding properties, it is understandable why the cultivation of spider silk has been of interest to many involved in material science.

However, unlike the cultivation of silk worms to harvest silk fibers, spider farming has not been proven successful because spiders are very territorial and cannibalistic. They do not tolerate being in close quarters with other spiders. Because of this, artificial, imitation spider silk fibers are being developed. Imitation spider silk is manufactured by a different process than other synthetic fibers. Fiber scientists try to imitate and use the natural ingredients, such as spider silk protein, used in real spider silk fibers, resulting in a biodegradable fiber.

A number of developmental scientists have embarked on this mission to create imitation spider silk fibers. Only two of these will be mentioned here. Bolt Threads, a California-based bioengineering company, has developed an imitation spider silk called Microsilk™ fibers and has begun using them in apparel products (see Figure 2.41). This imitation silk is not manufactured like other synthetic products, which are normally petroleum-based. This imitation spider silk is a protein-based fiber made by a fermentation process similar to brewing beer. The ingredients consist of yeast, sugar, salts, and water resulting in a biodegradable fiber and categorized as part of biotechnology. There are a number of companies trying to imitate spider silk protein to create the fibers. The fibers go through an extrusion process using a spinneret and therefore fiber shape can be manipulated just like other synthetic polymers extruded through a spinneret (see Figure 2.42).

Figure 2.41 Imitation spider silk, Microsilk™ fibers are used in apparel products.
Source: Courtesy of Bolt Threads, Emeryville, California.

Figure 2.42 Microsilk™ imitation spider silk fibers extruded through a spinneret.
Source: Courtesy of Bolt Threads, Emeryville, California.

Spiber® Technologies AB is a research company based in Stockholm that has modified and synthesized spider silk proteins by inserting the protein into the bacterium *Escherichia coli*, which in turn produces large quantities of the proteins. The end product can be used in medical textiles more specifically in regenerative medicine for tissue scaffolds and wound dressing (see Figure 2.43).

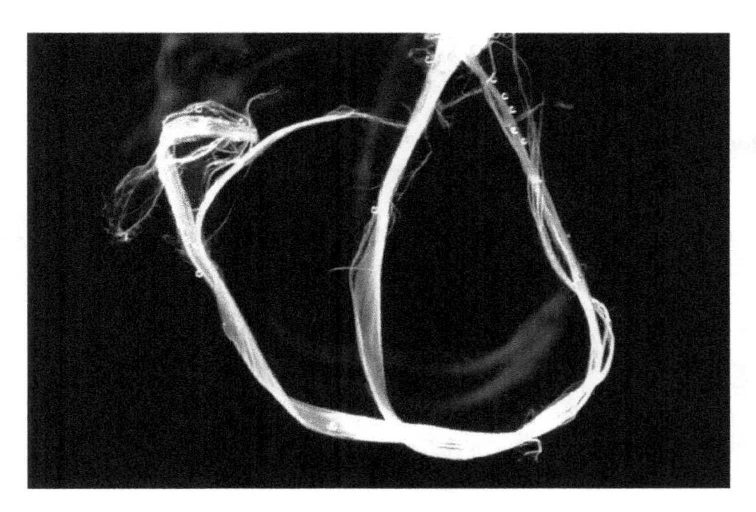

Figure 2.43 Synthesized spider silk fibers. Spiber fibers can be used in medical textiles, for example for tissue scaffolds and wound dressing. *Source:* Courtesy of Spiber® Technology AB, AlbaNova University Center, Stockholm, Sweden.

References

1 Kadolph, S.J. and Langford, A.L. (2002). *Textiles*, 9e. Upper Saddle River, NJ: Pearson Education.

2 Wildman, A.B. (1954). *The Microscopy of Textile Fibers*. Bradford, London: Wool Industries Research Association, Lund Humphries.

3 Petraco, N. and Kubic, T. (2003). *Color Atlas and Manual Microscopy for Criminalists, Chemists, and Conservators*. CRC Press, Taylor & Francis Group.

4 Saferstain, R. (2015). *Criminalists: An Introduction to Forensic Science*. Book Pearson Publishing.

5 Phan, K. H., Wortmann, F. J., Wortman, G. and Arns, W. (1988). Characterization of specialty fibers by scanning electron microscopy. *Proceedings from 1st International Symposium on Specialty Animal Fibers*, Aachen. Schrift. der Deutsches Wollforschungsinstitutet 103, 137–162.

6 Tortola, P.G. (1992). *Understanding Textiles*, 4e. New York, NY: Macmillan Publishing Company.

7 Von Bergen, W. and Krauss, W. (1942). *Textile Fiber Atlas: A Collection of Photomicrographs of Common Textile Fibers*. New York, NY: American Wool Handbook Company, Barnes Printing Company.

8 Lewis, D.M. and Rippon, J.A. (2013). *The Coloration of Wool and Other Keratin Fibers*. UK: Wiley.

9 Xian-Jun, S. and Wei-Dong, Y. (2008). Identification of animal fiber based on scale shape. In: *Conference Proceedings: Image and Signal Processing*

Conference (CISP'08), Sanya, Hainan, China (27–30 May 2008), vol. 3. IEEE https://doi.org/10.1109/CISP.2008.252.

10 Gordon, H.B. (1932). The identification of fibers. The hair fibers. *The Melliand Textile Monthly* 3 (11): 908–910.

11 Barber, E.J.W. (1991). *Prehistoric Textiles*. Princeton, NJ: Princeton University Press.

12 Hunter, L. and Mandela, N. (2012). Mohair, cashmere and other animal hair fibres. In: *Handbook of Natural Fibres*, Types, Properties and Factors Affecting Breeding and Cultivation, vol. 1 (ed. Y. Kozlowski), 196–290. Cambridge: Woodhead Publishing.

13 Vineis, C., Aluigi, A., and Tonin, C. (2011). Outstanding traits and thermal behavior for the identification of specialty animal fibers. *Textile Research Journal* 81 (3): 264–272.

14 Tonin, C., Bianchetto, M., and Vineis, C. (2002). Differentiating fine hairs from wild and domestic species: investigations of Shahtoosh, yangir, and cashmere fibers. *Textile Research Journal* 72 (8): 701–705.

15 Dobb, M.G., Johnson, F.R., Nott, J.A. et al. (1961). Morphology of the cuticle layer in wool fibers and other animal hairs. *Journal of the Textile Institute* 52 (4): 153–170.

16 Hunter, L. (1993). *Mohair: A Review of Its Properties, Processing and Applications*. Manchester: CSIR Division of Textile Technology, Port Elizabeth, International Mohair Association and the Textile Institute.

17 Langley, K.D. and Kennedy, T.A. (1981). The identification of specialty fibers. *Textile Research Journal* 51 (11): 703–709.

18 Skinkle, J.H. (1936). The determination of wool and mohair by scale and diameter. In: *Proceedings of the American Association of Textile Chemists and Colorists*, Lowell Textile School Massachusetts (November 16, 1936), 620–621. American Dyestuff Reporter.

19 Atav, R. and Turkmen, F. (2015). Investigation of the dyeing characteristics of alpaca fibers (Huacaya and Suri) in comparison with wool. *Textile Research Journal* 85 (13): 1331–1339.

20 Donn, T. and Yates, B.C. (2002). *Identification Guidelines for Shahtoosh & Pashmina*. Ashland, OR: National Fish and Wildlife Laboratory.

21 Tonin, C., Vineis, C., Bianchetto, M., and Festa-Bianchet, M. (2002). Differentiating fine hairs from wild and domestic species: investigations of Shahtoosh, yangir, and cashmere fibers. *Textile Research Journal* 72 (8): 701–705.

22 Langley, K. D. (1997). Shahtoosh fibers. The James Hutton Institute. http://macaulay.webarchive.hutton.ac.uk/europeanfibre/effnnew2kenneth.htm (accessed 18 August 2016).

23 Wortmann, F.J. and Arns, W. (1986). Quantitative fiber mixture analysis by scanning electrone microscopy. Part I: Blends of mohair and cashmere with sheep's wool. *Textile Research Journal* 56 (7): 442–446.

24 Wortmann, F.J. (1991). Quantitative fiber mixture analysis by scanning electron-microscopy. Round trial results on Mohair wool blends. *Textile Research Journal* 61 (7): 371–374.

25 Robertson, J. and Grieve, M. (2003). *Forensic Examination of Fibers*. London: Taylor and Francis.

26 Hearle, J.W.S. and Peters, R.H. (1963). *Fiber Structure*. London: Butterworth & Co. Ltd., Textile Institute.

27 Shen, Y., Johnson, M.A., and Martin, D.C. (1998). Microstructural characterization of bombyx mori silk fibers. *Macromolecules* 31 (25): 8857–8864.

28 Johnson, I., Cohen, A.C., and Sarkar, A.K. (2015). *J. J. Pizzuto's Fabric Science*, 11e. New York, NY: Bloomsbury Publishing.

3

Fur Fibers

3.1 Animal Fibers

The use of animal pelts or fur as clothing dates far back to ancient times. As early as 2000 CE, people in the Far East used animal fur not only as protection from the cold but also as luxury items. Indeed, throughout history, only the affluent could afford to wear fur. In addition to protection from the cold and display as a luxury item, animal fur has had other purposes. The ancient Egyptians believed that certain animals had magical powers and that wearing the fur of that animal transferred the power of the animal to a person. For example, in Egypt around 3000–300 CE, only the priests of a higher rank and kings could wear leopard skin during certain ceremonies during which it was believed the power of the leopard was transferred to them [1]. Leopard fur was considered to be so powerful that imitations were made by painting leopard spots on other fabrics; just as faux fur is made today.

Furs were prized by many dominant civilizations, such as the Romans, and it is believed that it was the Crusaders [2] who brought fur to Europe from the Far East along with other exotic items such as spices, perfume oils, and rugs. It is only in the last century that animal fur items have been available to the masses. When merchants began to gain more wealth during the Middle Ages, they began wearing the furs of ermine, mink, chinchilla, and sable. This offended the royalty who put forth sumptuary laws prohibiting the use of fur items by anyone other than royalty. Items made of animal pelts retained their popularity until the end of the twentieth century (i.e. the 1990s). Therefore, there are numerous historical pieces made solely or partly out of natural fur pelts. Only recently have animal activists been fighting against the use of real fur in the fashion industry. Also, more and more consumers realize that to kill an animal for fashion is not ethical. There have been some measures taken to combat animal cruelty. Animals are farmed for their fur, and animal activists demand that fur-farming practices be regulated. In some countries, Austria, the United Kingdom, and the Netherlands, for example animal farming is

Textile Fiber Microscopy: A Practical Approach, First Edition. Ivana Markova.
© 2019 John Wiley & Sons Ltd. Published 2019 by John Wiley & Sons Ltd.

banned altogether. Because of the alarming prevalence of illegal fur trade, organizations such as Convention on International Trade in Endangered Species (CITES), an international agreement of over 175 nations, work to protect endangered and threatened species in many countries.

Natural fur also differs in quality. The quality of fur depends on the animal's health condition, age, and diet. High-quality fur has a dense pile, with long, lustrous guard hair creating a fur coat which is soft and fluffy [3]. The fur can also differ based on the season. In winter, the fur increases in volume to keep the animals warm. Therefore, the fur will also differ in quality depending upon the season when the animal was killed.

Animal hairs may be simply grouped into two main categories as guard hairs and fur hairs. The guard hairs form the outer layer, also called the overcoat, of the hair, while fur hair forms the inner coat. Both hair types can be examined under a microscope on one slide, by placing them side by side using two cover slips, and there will be enough room for both on one slide. However, they should not be mounted in one mount because of the difference in thickness. As the guard hair is thick and underfur is light, you might be able to get only the thick hair in focus and not the under hair.

When viewing animal hairs under a microscope, it is important to note that the scale patterns on animal hairs are not consistent throughout the length of the hair shaft. Therefore, when the cuticle scale pattern is described in this text, it may mention scale pattern changes from the root of the fiber shaft to the center or base of the fiber shaft. The color of the hair can also change throughout the length of the fiber. To view the scale pattern, an examiner does not need more light, but less light. When viewing the medulla under a microscope, the examiner may benefit from more light.

Animal hair needs to be prepared for microscopic examination. Because animal hair contains oils, these oils need to be removed prior to viewing. The recommended cleaning solution is a combination of ether and ethanol/alcohol in a ratio of 1 : 1.

3.1.1 Scale Cast

Many experts recommend the use of a scale cast in order to view the entire length of the scale pattern. To make a cast of the fiber's surface, the hair is embedded in a soft medium such as a clear nail polish. When the medium is hardened, the hair is removed with tweezers leaving a distinct impression of the hair scale pattern. This specific impression, utilizing nail polish as the medium, can be viewed under a compound microscope. Scale casts are made with a variety of mediums. For students, nail polish, Duco cement, or simple clear tape are mediums that are easily accessible. Duco cement can be used instead of a clear nail polish, giving great results. Duco cement takes only five minutes to dry. Another simple method is to use a clear tape as a mounting medium by placing a hair fiber between a cover slip and a glass microscope slide held in place with a clear tape as a temporary mounting medium [4]. Transmitted light microscope is best for viewing the sample.

3.1.2 Cuticle Scales

The scales of animal hair have a variety of shapes and are categorized by their three most prominent shapes by some researchers: coronal, spinous, and imbricate [5].

1) Coronal (crown-like)
2) Spinous (petal-like)
3) Imbricate (flattened)

The first coronal shape has the appearance of crowns stacked on top of each other. There are variations within this category as some have more of a fold around the crown with pointy edges, and others have only onefold with dull edges. These uniquely appearing coronal scales are found in hair of small rodents and bats. They appear to be easily identifiable among the scales in hair categories. The second scale shape category is called spinous, and the scales are characterized by a petal-like appearance. These types of scales are found in mink, cat, and seal fur. The third general category of scale shapes is the imbricate shape, characterized by overlapping flattened scales exhibiting a narrow arrangement. This scale pattern may be found in animals and also in human hair.

The medulla is the central core of cells present in some animal hair. As the medulla is filled with air, when viewed with a microscope under a reflected light, it will appear as a white structure, whereas when viewed under a transmitted light, it will appear as a black, opaque structure. Most likely, a clear medium will be used when viewing animal fibers under a microscope, and a clear mounting medium can change the view of the medulla. When medium is used, the medulla will appear clear or translucent in transmitted light and almost invisible in reflected light. The best mounting medium to use to see the medulla is a mounting oil.

3.1.3 Rabbit, Hare, and Angora Rabbit Fibers

Garments made of animal fur can also be identified under the microscope. The majority of fur that is used in woolen clothing is chosen from the rabbit group, such as Angora rabbit, or other types of rabbit and hare fibers. Angora hair is fluffy compared to other rabbit hair fibers and is known for its softness. The fibers are lighter and warmer than wool fibers because of angora fiber's hollow core. Under the microscope, one can distinguish between rabbit and hare hair by the medullary cell arrangement. We have observed the medulla in hair fibers when examining wool fibers in Chapter 2. Now, the medullary cells may be arranged differently depending on the animal, and these differences in the arrangement are used as distinguishing characteristics. Thus, there are differences in the medullary cell arrangement between hare and rabbit hairs. Both rabbit and hare fibers have the so-called multiserial ladder medulla, but in the rabbit hair, the medullary rows are often interrupted and have some fused jointings with the rows besides them [6]. Other than that, the microscopic structure of common rabbit and hare fibers is essentially the same [7]. The sequence of scale pattern for both rabbit and hare is the same. The scale pattern is clearly marked by a double-chevron to single-chevron pattern in the middle of the shaft, which may change to a mosaic pattern toward the end of the fiber. In addition, sometimes, it changes from a double-chevron to an interrupted streak wave pattern [7].

On the other hand, the medullary rows of the hare hairs do not fuse together and most often appear as continuous rows [6].

The coats of fur-bearing animals have two main types of fibers, the outer layer and the undercoat [7]. The outer layer is usually referred to as guard hairs or the outer beard, whereas the undercoat is usually referred to as fur, down hair, or ground hair. The guard hairs are usually coarser than the finer and softer coat fur hairs. These two distinct types of hairs from the same animal appear different when viewed under a microscope. What is usually referred to as fur is typically recognized as the short undercoat fur hair. Guard hairs are straight shafts of hair that protrude through the undercoat fur hair layer (see Figure 3.1).

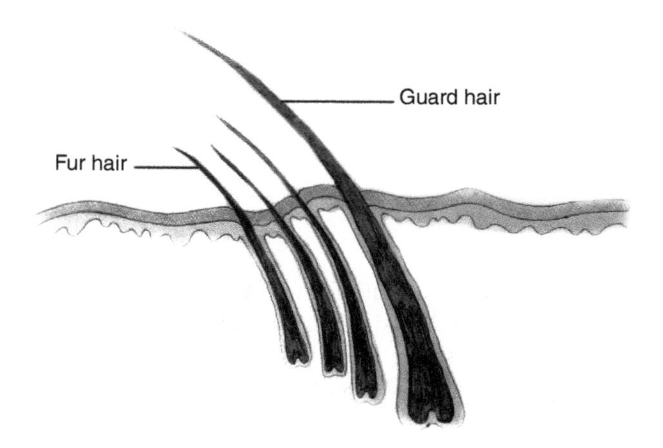

Figure 3.1 Animal hair types.

Even though the medulla of rabbit and hare fibers is said to be multiserial when identifying fur fibers, a uniserial ladder type of medulla may also be seen in finer fur fibers and multiserial when the fibers are coarse [7]. The coat of the hare is similar to that of the rabbit with the exception that most of the beard or shield fibers are longer and coarser.

The cross-sectional shape of rabbit and hare fibers is intriguing and very distinctive. The shape of guard fibers is not only elliptic but also bean-shaped, dumb-bell shaped [7], or ribbon-like [8]. The shape varies depending on the position along the length of the shield from which it has been taken [7]. The surface of cross-section has rows of medullary cells and air spaces [8].

3.1.4 Angora Hair

Angora hair that comes from Angora rabbits (*Oryctolagus cuniculus*) is considered a luxury fiber, also referred to as a specialty fiber. As it has a higher value, it is even more important to be able to distinguish it from other hair fibers. Some of its unique properties include fineness, lightness, durability, and luster. Angora rabbits originally came from Asia Minor, but today, they are raised in many other places: China, India, Europe, and the United States. Most Angora production takes place in China. Under the microscope, angora fibers have a hollow structure which is medullated. Hollow fiber structures usually increase the insulating ability because air gets trapped in the empty space inside the fibers and stays there producing a fiber which is also much lighter than other wool fibers. The longitudinal characteristic of angora hair fibers is somewhat different when compared to other rabbit or hare fibers. The scale shape characteristics and medulla type change are based on their position along the length of the fiber and based on the thickness of the fiber [7]. Although the medulla is believed to be one of the characteristics distinguishing angora hair from other rabbit hair as it is more uniserial, many researchers suggest that angora hair is not that different from other rabbit hair and is believed to have both ladder, uniserial and multiserial, medulla types [7]. Figures 3.2 and 3.3 illustrate Angora uniserial and multiserial medullae.

Angora hair fibers have different scale formations depending on the hair spot. At the base of the fiber, the scale pattern is irregular petal, or shallow irregular, waved mosaic or smooth. In midlength, the pattern is single or double chevron or double chevron, and on the tip of the fiber, it is single chevron [9] (see Figures 3.4–3.6). The interrupted streaked wave scale pattern, commonly seen in rabbit and hare fibers, is actually uncommon in angora hair fibers [7]. However, as previously mentioned, regarding scale characteristics, the same single- and double-chevron and waved, mosaic scale patterns also occur, as in the fibers of the hare and the gray rabbit, and at similar relative levels [7].

Because they are very fine and usually too fragile to be spun into its own yarn, angora fibers are blended with other fibers, such as wool to improve

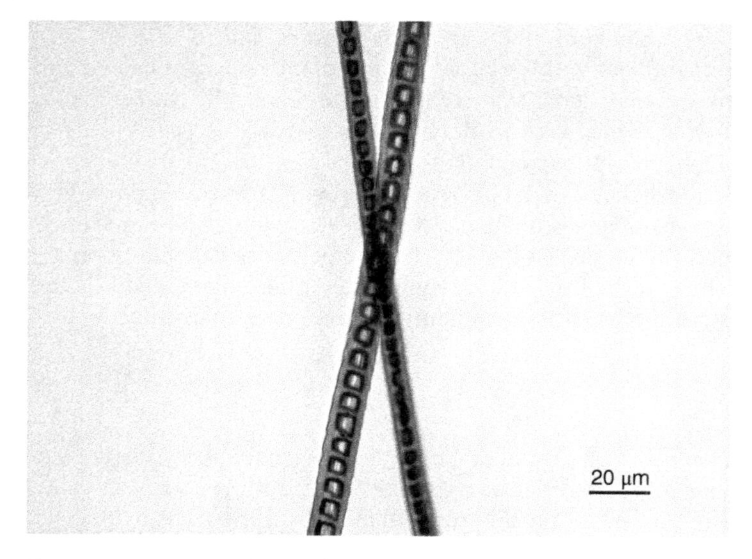

Figure 3.2 Longitudinal view of Angora rabbit hair depicting uniserial ladder medulla (compound microscopy).

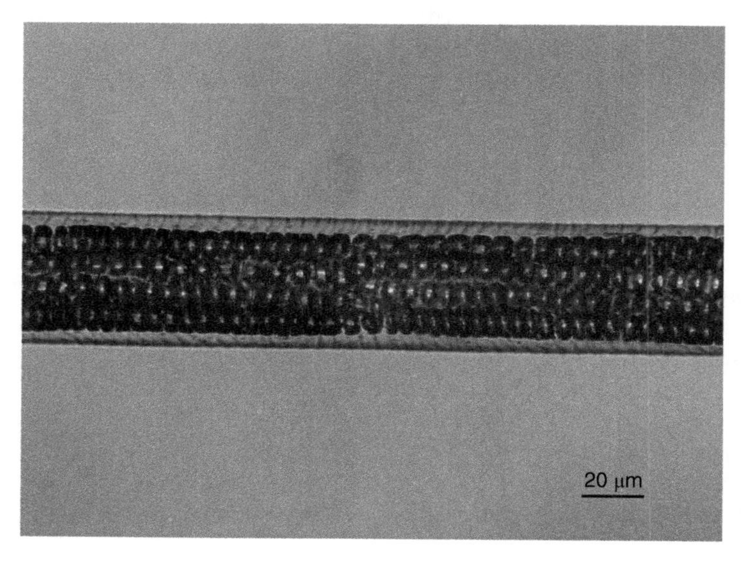

Figure 3.3 Longitudinal view of Angora rabbit hair depicting multiserial ladder medulla (compound microscopy).

performance. Thus, the fiber examiner must be careful when identifying the fibers under a microscope. When angora hair is blended with wool fibers, usually around 20% angora and 80% wool, an examiner will see two types of hair fibers – angora and wool. Some wool fibers also have a medulla but of a different

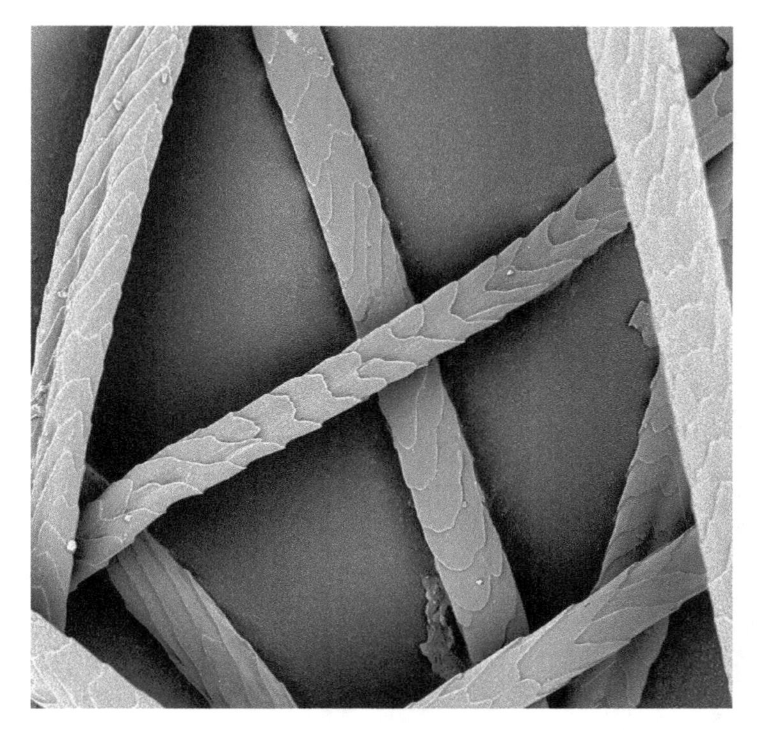

Figure 3.4 Longitudinal view of Angora rabbit fibers depicting variations of scale formations – from spinous (irregular petal) to chevron, and double-chevron type scale formations (1500×).

composition than the medulla in angora fibers. Angora fibers have medullary cells composition, whereas the medulla in wool fiber may vary but does not have medullary cells composition. In wool fibers, the size and shape of medulla varies greatly. It may consist of a continuous, interrupted, or fragmented line [6]. However, the medulla is seen mainly in coarse and medium wool fibers. More specifically, a reviewed older literature concluded that the medulla is only identifiable in coarse wool fibers with diameters larger than 35 μm [10]. Fine wools do not have a visible medulla. In cases when angora fibers are blended with fine wool fibers, there will be no medulla in the wool fibers to identify.

It is also by the cross-section that an examiner can distinguish angora fibers from other wool fibers as the cross-section of angora fibers contains a large hollow core, while other wool fibers do not, though both have a round to oval cross-section shape. Angora guard hairs have distinguishing dumb-bell shape (see Figures 3.7–3.9). Some kemp wool fibers have a medullary arrangement in the core of the fiber, which cannot be mistaken for a hollow core. Also, kemp wool fibers would probably not be blended with angora fibers. Angora fibers are more likely to be blended with other fine wool fibers.

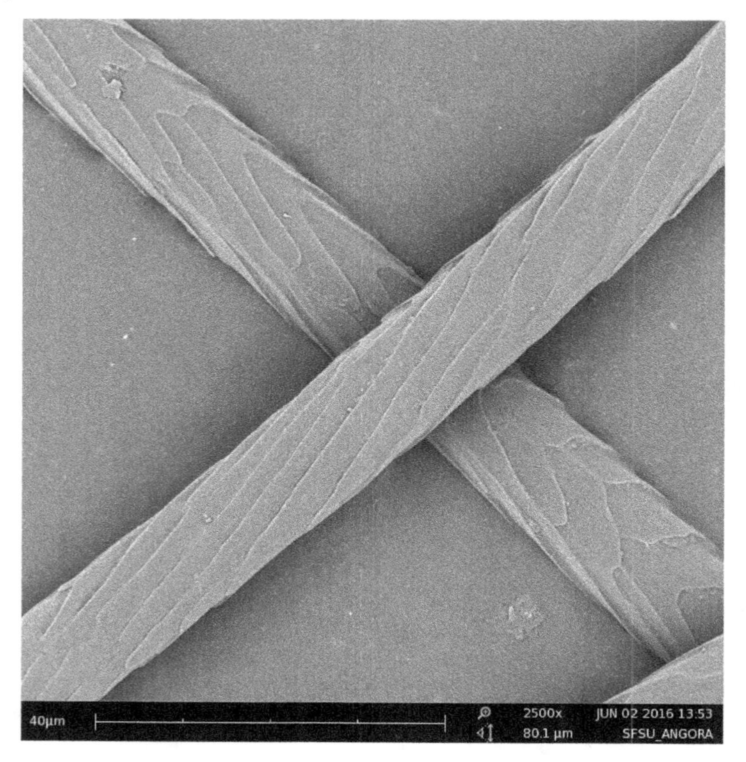

Figure 3.5 Magnified (2500×) view of Angora rabbit hair showing chevron-type scale formation.

3.2 Other Fur Fibers

As there are many pelts of animals used in clothing, their identification becomes crucial because they are sometimes sold under different names in order to deceive the customer. It is also important for fashion historians to properly identify the pieces in their collections. Today, faux fur is made out of synthetic fibers such as acrylic and polyester and surpasses the use of real fur. However, before synthetic fibers were available, most winter outdoor garments were made of animal fur.

Silver fox, chinchilla, mink, Persian lamb, and nutria are commonly raised on ranches. Chinchilla, mink, sable, platina fox, and ermine have always been very expensive. Although the United States has a Fur Products Labeling Act, mandating that garments containing fur be properly labeled, many furs are still dyed to make less-expensive furs appear more expensive. For example, muskrat may be dyed to resemble seal, and rabbit may be stenciled to look spotted. Furs are also dyed to improve their natural color or to give them unnatural colors:

Figure 3.6 Magnified (2000×) view of Angora rabbit hair showing spinous (petal-like) scale formations.

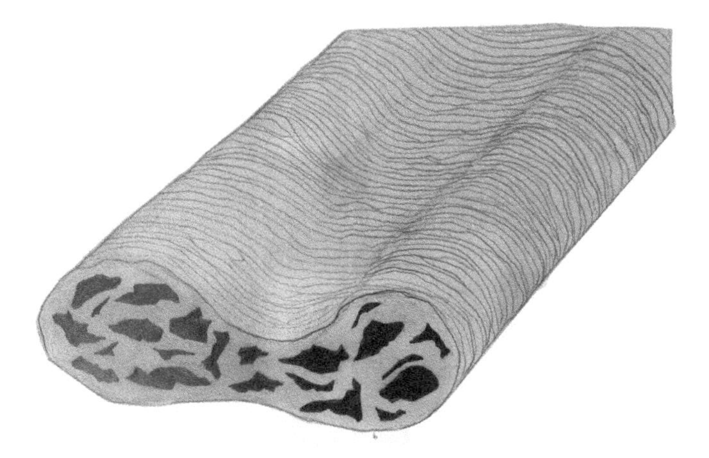

Figure 3.7 Structure of angora guard hairs with distinguishing dumb-bell shape.

Figure 3.8 Cross-sectional shape of angora hair showing variations of cross-sections, which include round, elliptical, and dumb-bell shapes. All the shapes have a hollow core, which is a specific characteristic to angora hairs.

red, green, etc. Some of the animal fur items, ermine, mink, and red fox, will be discussed in the following Sections 3.2.1, 3.2.2, and 3.2.4.

3.2.1 Mink and Ermine

Mink is a water-loving animal. There are a variety of mink pelts originating in Canada, North America, China, and also other Asiatic minks [7]. Mink fur was considered prestigious and was very pricey.

Mink and ermine hairs have a somewhat similar scale shape as they are coronal – dentate. However, these two fibers may be distinguished by their medullae as ermine hairs have more of a ladder shape medullae, while mink hairs have more of a globular-looking shape medullae. The globular and ladder shapes are very similar, and many experts identify mink fibers' medulla as ladder type. The medullae in mink fibers may gradually change to have a lattice filling as the hair expands. There is a difference between the medullae of coarse

Figure 3.9 Magnified (2500×) view of angora hairs showing hollow core cross-sections of different varieties including round, elliptical, and dumb-bell shapes.

and fine mink hair. Coarse hairs have a wide, broad gas-filled medulla taking up the majority of the fiber shaft, while fine hairs generally have a ladder type medulla [7]. The cross-section of American mink guard fibers is oval shaped with pigmented granules concentrated around the medulla, creating concentric circles [7]. The cross-sectional shape of fur mink fibers tends to be circular unlike those of other fur fibers such as rabbit's hair or otter's hair that tend to be polygonal [7].

In addition, the cuticle scales will change; from the root of the hair, they have a coronal dentate shape, which then becomes an imbricate extreme crenate shape for the thicker part of the hair. However, the scales may also have a spinous almost elongated petal shape through the fine portion of the hairs. American mink hair is also described as having a diamond petal to a coarse pectinate scale pattern in the shaft portion of the guard hair [7]. However, the scale pattern of fur hairs is described as being lanceolate to pectinate through the entire length of the fiber. The hair is uneven in diameter, of a so-called spear shape, and colored darker brown.

3.2.2 Kolinsky Mink

The finest species of an Asiatic mink is called kolinsky mink (*Mustela lutreola* sibirica). Kolinsky hairs were formerly used in manufacturing small brushes. The medulla is a ladder type [7] of fine fibers and a unicellular ladder type of coarse fibers. The cross-sectional shape of guard hairs is similar to that of American mink hair in that it is oval but without a pigmented cortex. The fine hairs, on the other hand, have a more rectangular shape in cross-section [7]. The cuticle scale pattern differs from guard hairs to fine hairs. Guard hairs near the root have a lanceolate diamond petal pattern, which further up the fiber changes to an irregularly waved mosaic pattern. Fine fibers' cuticle scales may be described as a diamond petal coarse pectinate pattern, which will also change along the fiber into irregular-petal pattern [7].

Chinchilla hairs appear to have a scale pattern that is simple coronal to dentate coronal somewhat similar to that seen in mink and ermine but with scales that are denser and close together. The shape of the medulla is uniserial ladder, which may change to biserial or aeroform in the thicker regions.

3.2.3 Raccoon Dog

Raccoon hair has a medulla similar to that of dog hair as it is amorphous through the basal region and may later change to vacuolated. It is dark and broad, but does not appear as broad as a dog hair medulla. In raccoon hair, the medulla appears to be less even than that of dog hair as it is somewhat more vacuolated. The cuticle scale differs from dog hair as they are more of a diamond-shaped petal in basal and may gradually change to imbricate.

Raccoon dog (*Nyctereutes*), a different species from a raccoon, has been also called a Russian raccoon [11]. Raccoon dog hair is sometimes used today in fur coats, but it is unfortunately mislabeled. Therefore, it is important to have a source of fiber comparison. Raccoon dog's hair is longer than raccoon's hair, but they have similar hair structure and appearance [11].

Raccoon dog hair also shows diamond-shaped petal scale structure (see Figure 3.10) in basal and may gradually change to imbricate (Figure 3.11). These imbricate fibers have scale margins that have a rippled structure, meaning the scale margins have deeper indentations [7] (see Figure 3.12). Raccoon dog's fur (under hair) and guard hair depicts substantial differences from guard hair in size and in scale structure (see Figures 3.13 and 3.14). Medulla of a raccoon dog's guard hair is partially vacuolated and is of the amorphous type (see Figure 3.15), whereas the medulla of under hairs is uniserial ladder type (see Figure 3.16). Additionally, raccoon dogs' hair may be differentiated from other fibers by numerous melanin granules, generally spherical, throughout the cortical cells [12]. These granules can be seen after surface-etching under high magnification using scanning electron microscopy (SEM) of 1 μm section of the fiber. Even though raccoon dogs' hair is not desirable in consumer goods

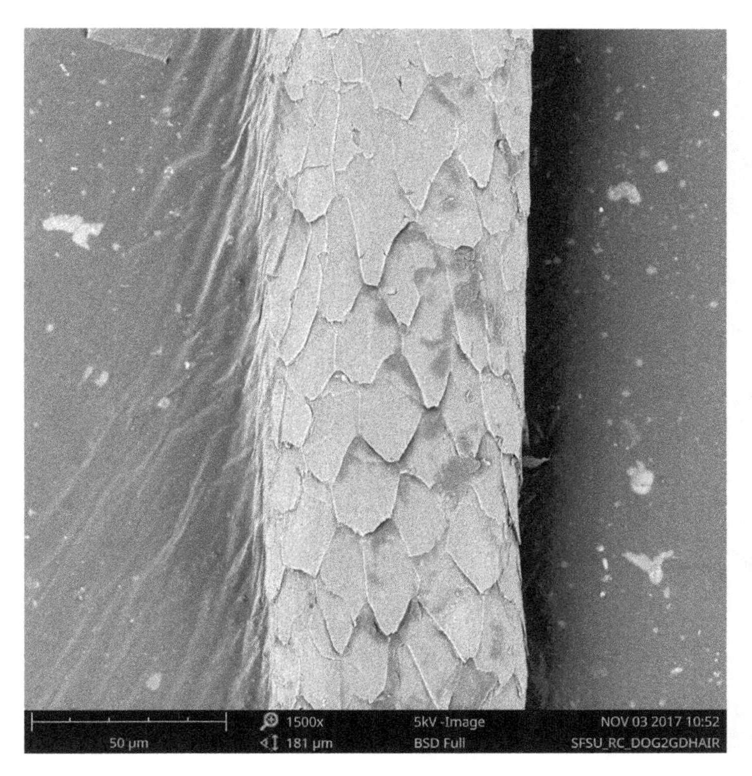

Figure 3.10 Longitudinal view of Raccoon dog hair fibers showing diamond-shaped petal scale structure (1500×).

today, clothing historians may encounter some items containing raccoon hair as many men's coats were made of it in the 1920s (Figure 3.17).

3.2.4 Red Fox

Red fox hair has a slightly elongated diamond petal-shaped cuticle scale pattern [2]. The scale pattern may change to a mosaic pattern along the length of the fibers. The medulla varies from a uniserial ladder type [2] to a vacuolated type. The color of the hair will vary from a red to brown tip with a brown, gray to yellow base. In the past, hare fur fibers have been dyed and sold as red fox hair [2].

3.3 Faux Fur

The fake fur manufacture began due to the high prices of real, genuine fur products. However, today, consumers prefer faux fur because of animal cruelty issues. Fur products are sometimes purposely mislabeled to mislead

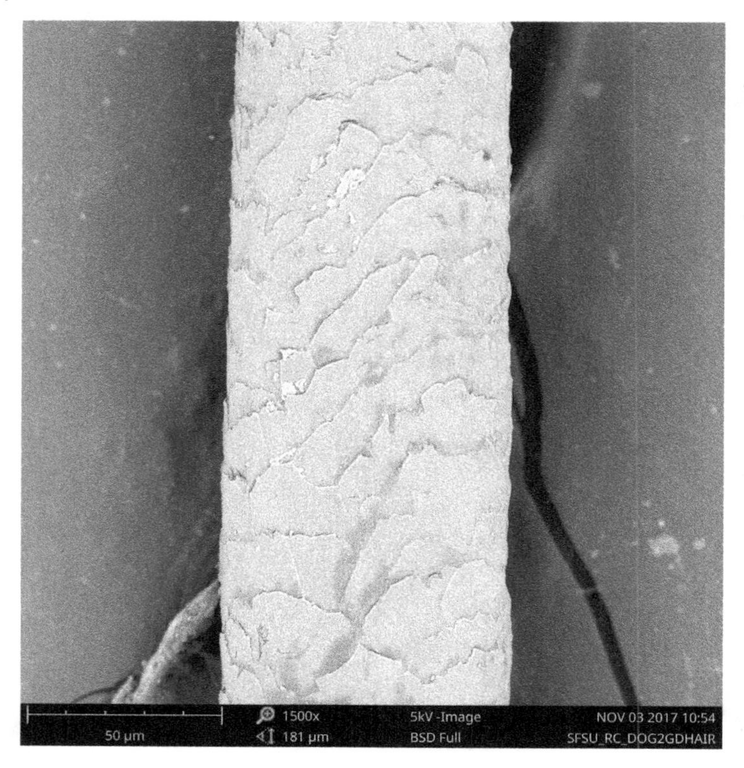

Figure 3.11 Longitudinal view of Raccoon dog hair fibers showing imbricate scale structure (1500×).

consumers into believing that they are purchasing faux fur garment when they are in fact getting real animal fur. This clandestine substitution had been practiced in the United States by retailers such as Neiman Marcus and Nordstrom. Consumers and retail buyers can be easily misled because of the difficulty in differentiating faux fur from real fur by merely touching it. The quality and resemblance of faux to real fur have increased incredibly by making the color and softness of hairs almost identical to that of true fur; they are nearly indistinguishable. However, in using only a simple light microscope with low 10× magnification, the differences between faux fur and real fur are obvious and difficult to miss because basically you are differentiating a natural fiber from a synthetic fiber. As mentioned earlier, the first obvious difference is that natural fibers are more irregular in their shape than synthetic fibers, which are usually tubular with straight edges with an overall even shape. Acrylic will be discussed in more depth as most faux fur is made out of polyester, acrylic, or modacrylic. It is these three fibers one would identify when viewing faux fur under the microscope [3].

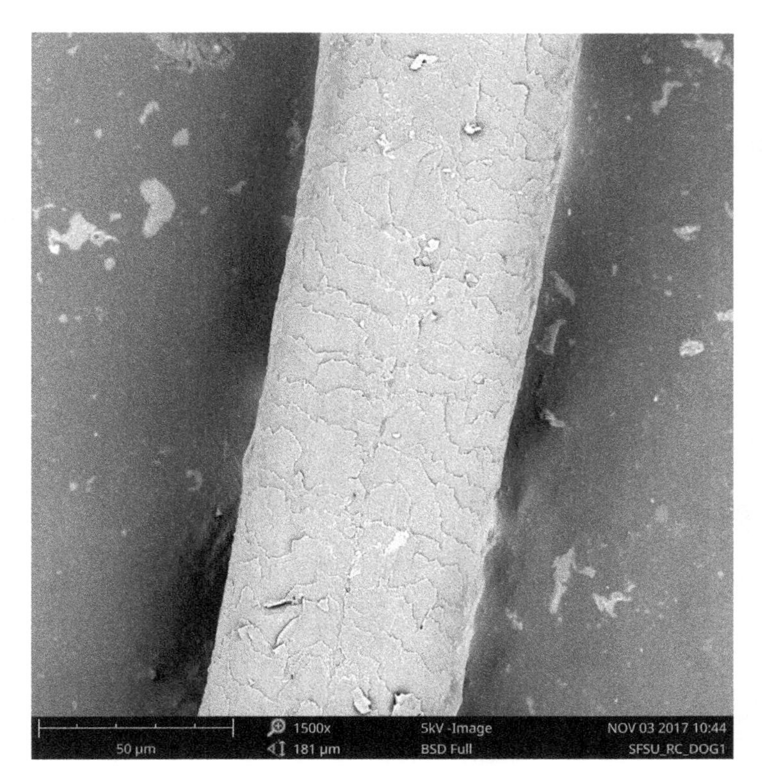

Figure 3.12 Longitudinal view of Raccoon dog hair fibers showing imbricate scale structure with scale margins having rippled structure – scale margins with deeper indentations (1500×). *Source:* From Wildman 1954 [7].

Among all other manufactured fibers, acrylic fibers imitate animal hairs, such as wool, most often [3]. Under a microscope, in the longitudinal view, acrylic fibers display some evidence of texture on the surface when viewed under a low 40× magnification via a light microscope. When the fibers are identified under a high enough magnification, for example 3000×, one will notice that this texture actually cracks running lengthwise on the surface of the fiber (see Figure 3.18). These cracks will become larger with wear and abrasion, and tiny parts of the crack/fiber will break off. This is the reason why clothing made of acrylic has a pilling problem. Some experts also describe the fiber surface of acrylic as pitted and irregular [3] and consider acrylic fibers to be more wavy rather than rod-like as other synthetic fibers such as polyester and nylon [13]. The most common cross-section shapes of acrylic fibers are round, bean, and dog-bone shaped (see Figure 3.19). However, acrylics may be produced with other special cross-sectional shapes such as ribbon or mushrooms [13].

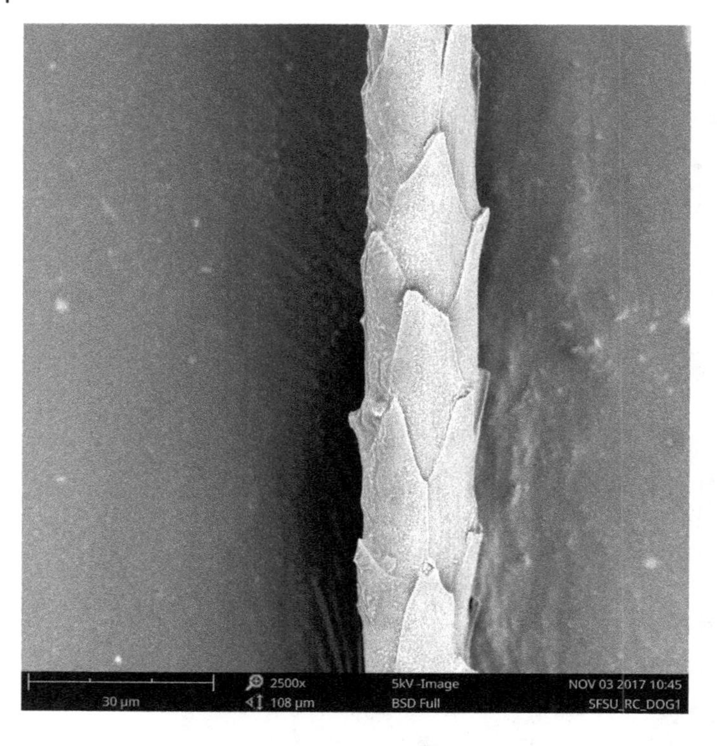

Figure 3.13 Longitudinal view of Raccoon dog fur fibers showing spinous (petal-like) scale shape (2500×).

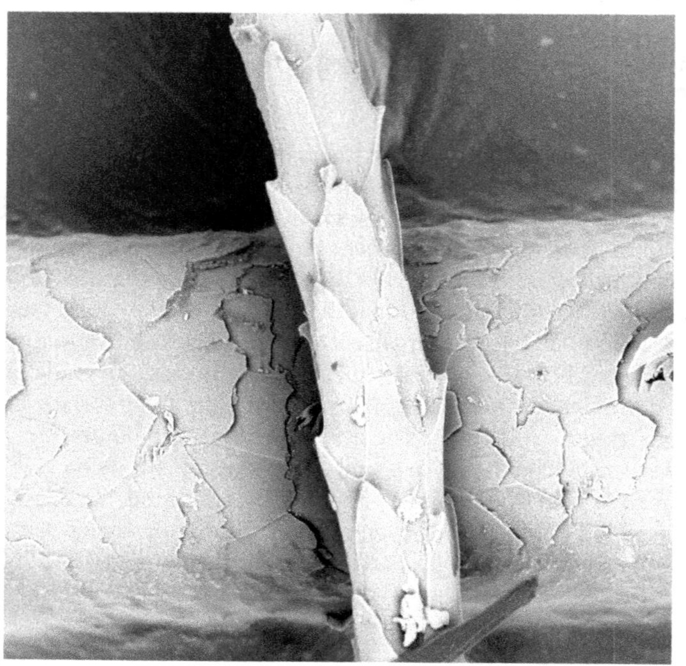

Figure 3.14 Size comparison of fur (fine) and guard (coarse) Raccoon dog hairs (2500×).

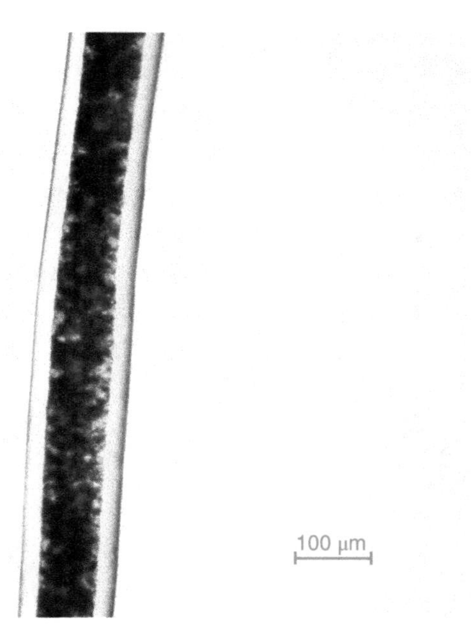

Figure 3.15 Medulla of Raccoon dog's guard hair appears to be partially vacuolated and amorphous (compound microscope).

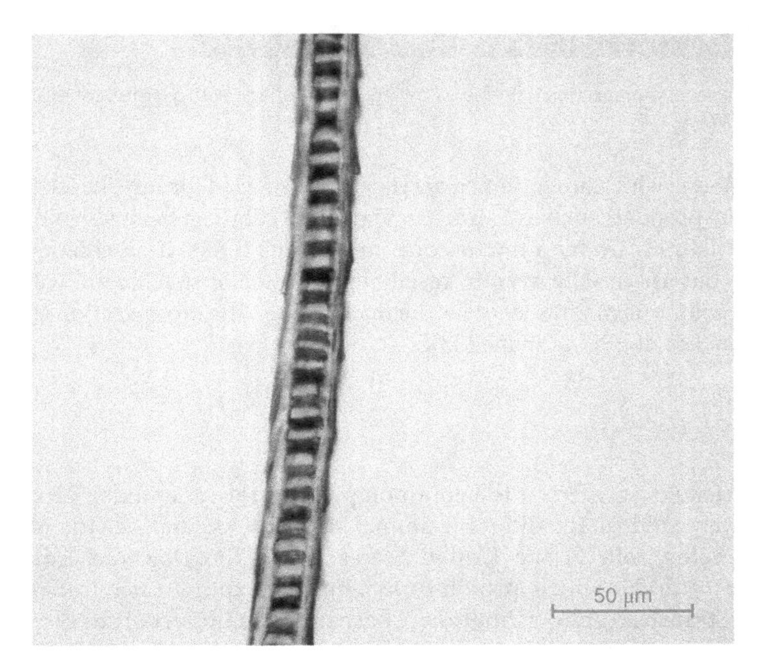

Figure 3.16 Medulla of Raccoon's dog under hair with appearance of uniserial ladder type medulla (compound microscope).

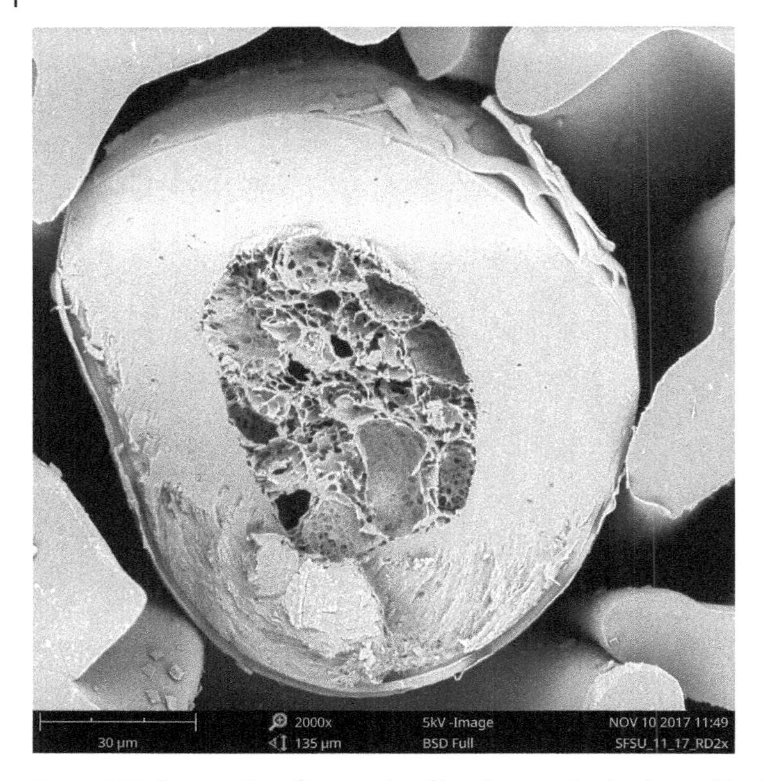

Figure 3.17 Cross-section of raccoon hair fiber showing circular shape with a large central canal opening (2000×).

Modacrylic fibers, which are modified acrylics, are also used for fur-like fabrics and hair-like products such as hairpieces and wigs. They can be made into a variety of fur fabrics. Under a microscope, modacrylic fibers are a creamy white color [3] but are usually dyed to resemble the color of specific animal hair; this is the color one might see under the microscope. The cross-section is irregularly shaped or dog-bone shaped [3].

3.4 Dog and Cat Fur

In order to mislead customers, it is a common practice for cat and dog furs to be deliberately sold under different animal names. Dog and cat fur is banned from being sold in the United States under the *Dog and Cat Protection Act of 2000*, which prohibits its import, export, sale, trade, advertisement, transport, or distribution. Therefore, it is important to distinguish cat and dog hair from other animal hairs. Only raccoon dog hair,

Figure 3.18 Longitudinal view of artificial fur (acrylic fibers) depicting textures surface or surface with cracks running lengthwise on fiber surface (2000×).

which may be difficult to identify, is legally allowed to be sold in the United States.

Dog hair usually does not have prominent scales. The scales are not only of a spinous shape, but also have a square-like shape. Dog's hair can be distinguished from other animal hair mainly by its root which has a distinctive spade-like shape. Dog hair has a very distinctive broad, dark medulla, characterized by ovoid bodies, solid structures of a spherical or oval shape, running down the fiber on its sides. It has also been described as amorphous or shapeless, when compared to other fibers. Some experts claim that a uniserial ladder medulla can also be seen in fine dog hairs. It has coarse pigment extending to the root. The diameter may vary from fine to coarse, usually coarser than cat hair, with a medium overall diameter of 75–150 μm. A major issue with dog hair is its high variability depending upon the breed of dog.

Cat hair has a fine diameter and is finer than dog hair. Cat fur hair has a uniserial ladder medulla similar to that of rabbits, though rabbits' coarse hair

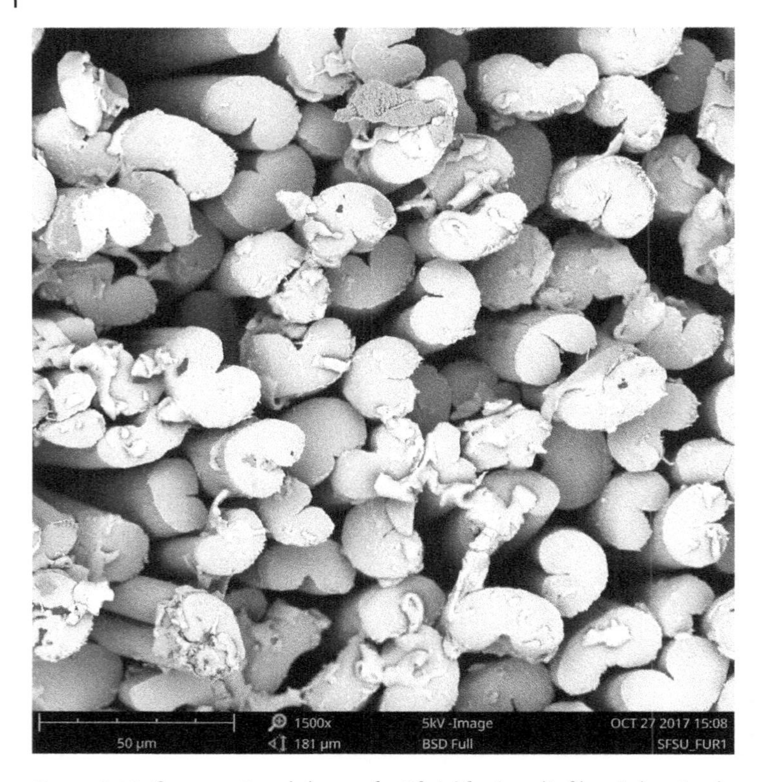

Figure 3.19 Cross-sectional shape of artificial fur (acrylic fibers) showing bean shapes fibers (2000×).

differs as it has a multiserial medulla [7]. Coarse cat hair has a more vacuolated medulla. Because the fur undercoat hairs of both cat and rabbit are similar, it is suggested that both fur undercoat hairs and guard hairs be examined as well. The scales of cat hair are very prominent and spinous. The spinous shape of the scales has short, round, thick, fat petals somewhat looking like a hair beard.

3.4.1 Karakul

Karakul (*Ovis*), sometimes referred to as Persian lamb, is a breed of sheep introduced to the United States in the nineteenth century. It is an ancient fiber, believed to date back to 1600 CE [14]. Karakul wool is mainly distinguished by its wavy curls, which, as the lamb matures, open and lose their pattern. When the lamb is born, it yields the best curls because they are tight, silky, and smooth. Coats and hats made of this type of wool were in fashion at the beginning of the twentieth century and were considered luxurious. However, they diminished in popularity due to pressures from animal protectionists and ethical considerations as the lambs had to be killed shortly after they were born to obtain the nicely curled wool. Sometimes lambs were killed while in womb to obtain this superior wool. This fleece was also referred to as astrakham. Today, fashion historians will encounter items from this type of wool in outer garments from the United States, Europe, Asia, and Russia.

This curly wool is also distinguishable by the color of its fleece, predominantly black, though other natural colors of the fiber varying from black, gray, and brown occur. The mean fiber diameter ranges from about 30 to 40 μm with a typical mean value of 35 μm [14]. Fine fur fibers are rare, 8–15 μm in diameter, and believed to be indistinguishable. The intermediate and guard hairs may be grouped together. The intermediate hair may be identified by the scale structure, characterized by simple coronal scales under 400× and as coronal dentate under 1000× magnification. The scales have a smooth surface and straight scale margins. Fine fibers do not have a medulla. The intermediate and guard hairs have a diameter of 40–80 μm with a circular cross-section, becoming elliptical toward upper portion of the hair shaft. The medulla of intermediate and guard hairs is wide, though narrowing toward the end of the hairs. It is an unbroken medulla type with an amorphous filling [15].

3.4.2 Optical Microscopy

Helpful tip for fiber examiners: clean animal fibers prior examination as animal fibers may have debris, which can be cleaned with distilled water or acetone.

Other helpful methods have been developed to help differentiate cat hair from dog hair. The first method is the medullary fraction (MF), which is the ratio of medulla width to hair width [16]. Optical microscopy should be sufficient in viewing the medulla. Some experts believe that hairs from cats and dogs have similar medullae, a ladder type in fine hairs, and a wide lattice type in coarse hairs. Cat and dog hairs appear to differ in their medulla width to fiber width ratio. Cat hair has wider medullae compared to those of dog hair [16].

The second method is the observation of fiber scale pattern progression, the sequence of different scale patterns along the body of the hair [16]. The surface architecture of hair fibers is best determined through the use of SEM because the scale formations may be seen from every angle [16]. In cat hair, the scale pattern is progressive in this order: starting from the root with irregular mosaic pattern that changes to petal, then to regular mosaic, remaining regular mosaic until the tip of the hair. In dog hair, the scale pattern progression varies and is very different and more complex.

The third method is the hair length. Although there are many dog and cat breeds, hair length may be used as a discriminating factor. Dog hair ranges from 0.6 cm for short-hair dogs to 15.3 cm for long-hair dogs. In comparison, cat hair length is more consistent, mainly ranging from 1.2 to 5.0 cm, with the exception of Persian and Siamese breeds that may have longer hair than the average cat [16].

3.4.3 Measuring Hair Length

To measure the length of hairs, light microscopy is adequate. The examiner will need some additional microscopic supplies for the task, an eyepiece reticle and a stage micrometer. An *eyepiece reticle* is a small piece of glass with a ruler or grid etched into it, which can be placed into the microscope eyepiece.

A *stage micrometer*, the second part needed, is basically a microscope slide with a ruler etched into it. Glass stage micrometers are suitable for transmitted light microscopy, while a metal stage micrometer is used for reflected light microscopy. A stage micrometer is used to calibrate the eyepiece reticle before making microscopic measurements.

References

1 Tortora, P.G. and Eubank, K. (2010). *Survey of Historic Costume: A History of Western Dress*, 5e. New York, NY: Bloomsbury Publishing.
2 Hausman, L.A. (1920). The microscopic identification of commercial fur hairs. *The Scientific Monthly* 10 (1): 70–78.
3 Kadolph, S.J. and Langford, A.L. (2002). *Textiles*, 9e. Pearson Education: Upper Saddle River, NJ.
4 Crocker, E.J. (1998). A new technique for the rapid simultaneous examination of medullae and cuticular patterns of hairs. *Microscope* 46 (3): 169–173.
5 Saferstain, R. (2015). *Criminalists: An Introduction to Forensic Science*. Harlow: Pearson Education Limited.
6 Von Bergen, W. and Krauss, W. (1942). *Textile Fiber Atlas: A Collection of Photomicrographs of Common Textile Fibers*. New York, NY: American Wool Handbook Company.
7 Wildman, A.B. (1954). *The Microscopy of Textile Fibers*. Bradford/London: Wool Industries Research Association/Lund Humphries.
8 Petraco, N. and Kubic, T. (2004). *Color Atlas and Manual of Microscopy for Criminalists, Chemists, and Conservators*, 1e. Boca Raton, FL: Taylor & Francis Group LLC/CRC Press.
9 Appleyard, H.M. (1960). *Guide to the Identification of Animal Fibers*. Wool Industries Research Association.
10 Rippon, J. (2013). The structure of wool. In: *The Coloration of Wool and Other Keratin Fibers* (ed. D.M. Lewis and J.A. Rippon), 1–35. Delhi: Wiley.
11 Hicks, J.W. (1981). *Microscopy of Hairs: A Practical Guide and Manual*. United States Department of Justice, Federal Bureau of Investigation.

12 Sato, H., Miyasaka, S., Yoshino, M., and Seta, S. (1982). Morphological comparison of the cross-section of the human and animal hair shafts by scanning electron microscopy. *Scanning Electron Microscopy* 1: 115–125.
13 NPTEL (2014). Online lectures website. Acrylic fibers. www.nptel.ac.in/courses/116102026/39 (accessed 18 July 2017).
14 Hunter, L. and Mandela, N. (2012). Mohair, cashmere and other animal hair fibres. In: *Handbook of Natural Fibres*, Types, Properties and Factors Affecting Breeding and Cultivation, vol. 1 (ed. Y. Kozlowski), 196–290. Cambridge: Woodhead Publishing.
15 Galatík, A., Galatík, J., and Krul, Z. (1997). Furskin identification. Microscopic atlas. www.furskin.cz/overview.php?furskin=Ovis%20aries%20-%20Semi-Persian%20lamb%20skin (accessed 13 September 2017).
16 Peabody, A.J., Oxborough, R.J., Cage, P.E., and Evett, I.W. (1983). The discrimination of cat and dog hairs. *Journal of the Forensic Science Society* 23: 121–129.

4

Regenerated Cellulosic and Protein Fibers

4.1 Regenerated Cellulosic Fibers

Rayon is a manufactured regenerated cellulosic fiber. It differs from other manufactured or synthetic fibers due to its cellulosic content. The production of manufactured fibers was motivated by the desire to make a fiber closely resembling silk. Unsuccessful attempts to imitate silk date back to 1664. It was only in the late nineteenth century that these attempts were successful.

Rayon, the first manufactured fiber, was invented by an English scientist, Charles Frederick Cross, and his colleagues. They invented the viscose process, used to manufacture viscose rayon, in 1891. Most rayons today, including bamboo rayon, are manufactured using this viscose process. Rayons are made out of cellulose, a natural compound, taken from a variety of plant sources; one example is wood pulp harvested from eucalyptus trees, which is later mixed with other chemicals. Under the microscope, we cannot see cellulosic content because in the manufacturing process, the natural cellulose is converted to a soluble cellulosic derivative. The resulting cellulosic solution is then extruded through a spinneret, a device resembling a showerhead, into a chemical bath where it hardens into fibers. This is called wet-spinning coagulation, which is used in manufacturing fibers. Thus, the shape of rayon fibers or any other manufactured fibers, examined microscopically, is a result of the spinneret's shape. The spinnerets are engineered to produce the desired fiber properties.[1]

4.1.1 Viscose Rayon

Different types of rayon possess some distinguishing characteristics when viewed under a microscope [1] (see Figure 4.1). Viscose rayon, the first manufactured cellulosic fiber, is striated in the direction of its axis (see Figure 4.2).

1 Because rayon fibers are made of a natural polymer, cellulose, the fiber/fabric properties are those of other natural fibers – good absorbency, breathability, and soft hand.

Textile Fiber Microscopy: A Practical Approach, First Edition. Ivana Markova.
© 2019 John Wiley & Sons Ltd. Published 2019 by John Wiley & Sons Ltd.

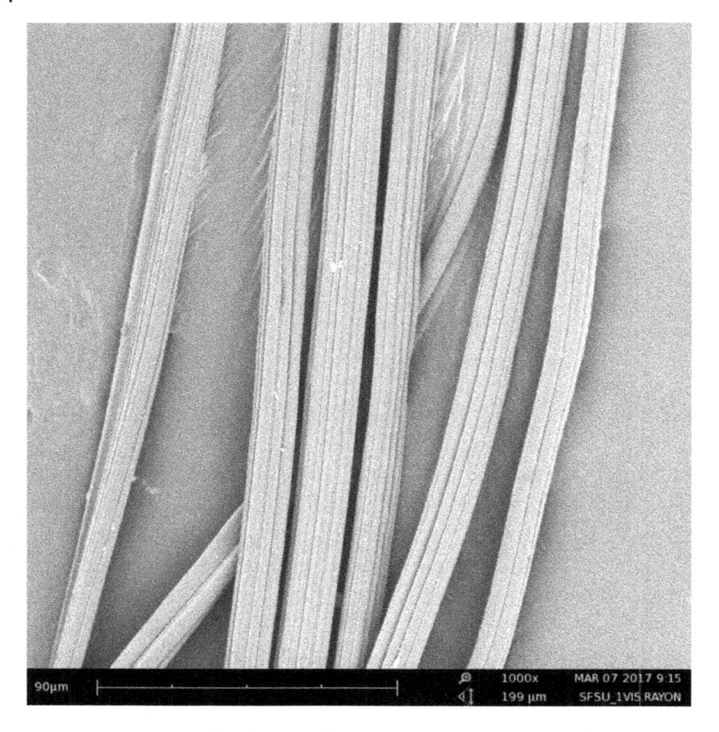

Figure 4.1 Longitudinal view of viscose rayon showing its distinguishing characteristics – striations in the direction of its axis (1000×).

It has been pointed out that viscose rayon is different in appearance from any other type of fiber and should be immediately recognizable [2]. Viscose rayon has been described as a fiber that is deeply grooved longitudinally with an irregular cross-section [2]. The cross-section of viscose rayon's fibers is irregular, toothed, and has voids that make the fiber absorbent [1]. The serrated cross-section of viscose rayon is the result of coagulation. During this stage of the fiber-making process, the fiber, having been extruded through the spinneret, is placed in an acid bath where it reacts by shrinking slightly and creating the serrated edges (see Figure 4.3). The serrated shape of the cross-section results in striations along the length of the fiber, as the fiber has lost a part of its solvent during the process. Note that students sometimes misidentify unseparated, smooth filament fibers for striations when viewing them under a light microscope. To avoid misidentification, students must be careful to completely separate fibers before placing them on microscope slides. It is interesting to note that these striations provide some advantage when dyeing fibers as they increase the fiber's surface area [3]. As rayon fibers are very bright, delustering pigments are usually added to the solvent to create duller finish for the fibers

Figure 4.2 Magnified view of viscose rayon fibers depicting deep grooves also called striations (1500×).

[3]. These pigments, easily seen under a light microscope as cross-sectional and longitudinal views, appear as dark specs or dots evenly distributed throughout the fibers and are usually referred to as granules. The chapter on synthetic fibers provides more information on fiber delustering.

4.1.2 Bamboo Rayon

Bamboo rayon fibers are obtained from a bamboo species known as Moso bamboo (*Phyllostachys heterocycla pubescens*), which comes from China. Regenerated cellulosic bamboo fiber was first manufactured in 2002 [4]. Even though it has only been recently developed, it still uses the less desirable viscose production method and thus is not considered sustainable as has been claimed. This is also why it is called bamboo viscose fiber. Bamboo rayon under the microscope is not round and smooth, but is similar to that of viscose fibers that are striated in the direction of the fiber axis (see Figure 4.4). Cross-section of bamboo rayon has a serrated shape similar to viscose rayon (see Figure 4.5). Bamboo fibers are described as having many microfiber surface grooves,

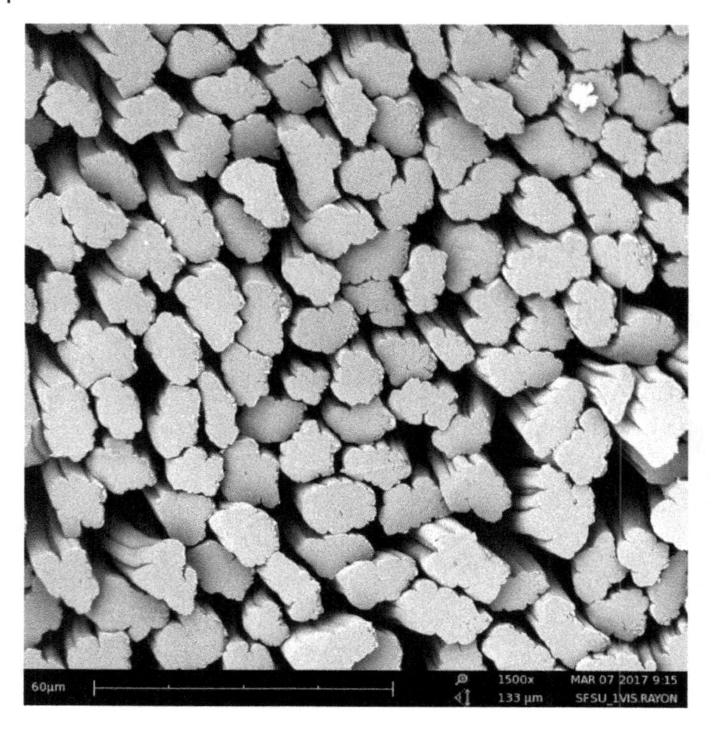

Figure 4.3 Cross-sectional view of viscose rayon depicting serrated cross-section with voids – voids make the fiber absorbent (1500×). *Source:* Reprinted with permission from Xu et al. [1]. Copyright © 2007, Wolters Kluwer.

a horizontal irregular oval and round waist, with the cavity filled with circumferential and small gaps [5]. Because of the micro gaps or tiny, capillary-like holes, bamboo clothing provides immediate moisture absorption and evaporation [4, 5]. Regenerated bamboo fibers provide many advantages for the wearer that cannot be found in other fibers, including natural antibacterial and UV protective characteristics. For this reason, garments from bamboo fibers are used in children's clothing and are preferred for pregnant women. Because bamboo fibers are also soft and breathable, they are used in sport clothes and underwear products [1].

4.1.3 High Wet Modulus (HWM) Rayon

High wet modulus (HWM) or modal is a second-generation rayon also made using the viscose process. It has properties similar to viscose rayon, such as comfortable and soft, but it also has a high wet strength as implied by its name. The end fabric is the most similar to cotton, but it has a more

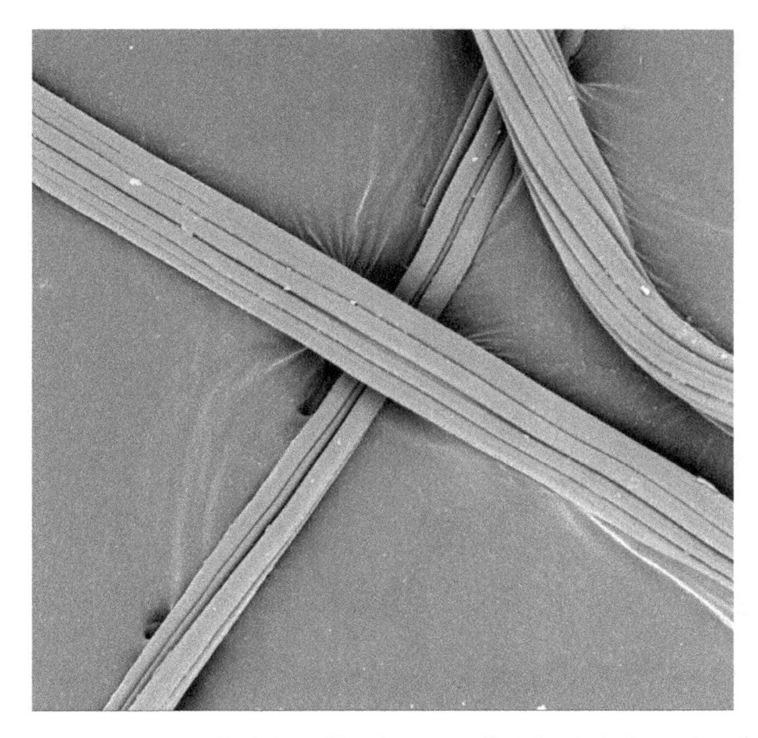

Figure 4.4 Longitudinal view of bamboo rayon fibers but is similar to that of viscose fibers which are striated in the direction of the fiber axis (1500×).

lustrous hand. Under microscopic examination, modal fibers are not as striated as viscose fibers, having few visible grooves or indentations. The cross-section of modal fibers appears slightly irregular, having a bean or peanut shape with multiple pores. The pores, appearing as gray shading, are small, increasing in size toward the center of the fiber cross-section [6]. In addition to having differently shaped cross-sections, modal, viscose, and lyocell rayons also differ in their pore distribution in cross-section of the fibers. Viscose rayon (star-like cross-section) has the most porous and spongy core with pores mainly situated around the center of the fiber and not on the irregular indentations. Whereas lyocell fibers (circular cross-section) are more homogenous in cross-section with only small pores visible [6], modal fibers have a rounder cross-section [3]. Modal rayon, even though made with the same viscose process as viscose rayon, does not have the serrated edges in cross-section because the acid bath is less concentrated during the coagulation process. HWM rayon retains its microfibrillar structure and therefore is more similar in performance to cotton fibers than to other rayon fibers (Figures 4.6 and 4.7) [3].

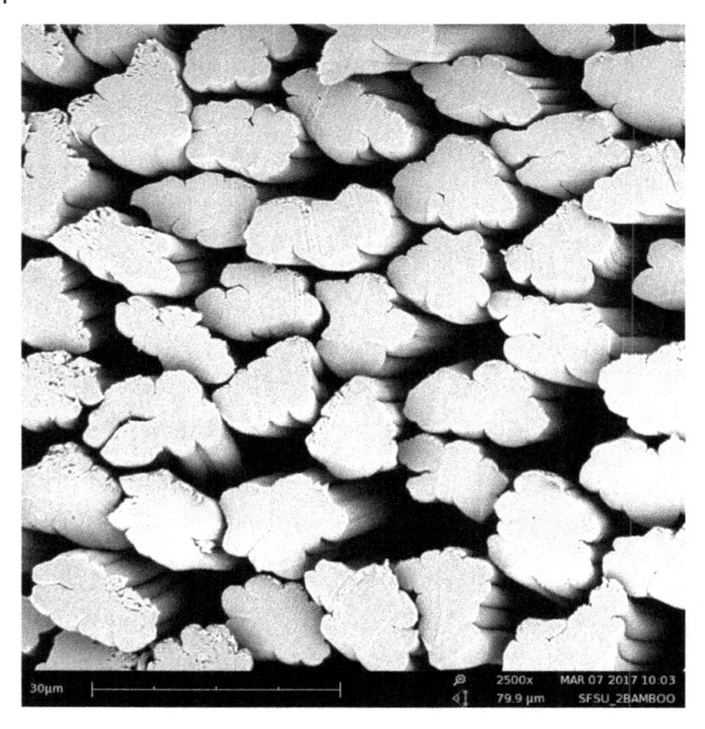

Figure 4.5 Cross-sectional view of bamboo rayon fibers. Bamboo rayon has a serrated shape similar to viscose rayon (2500×).

4.1.4 Cuprammonium Rayon

Cupra rayon, the rayon that most resembles silk, was also invented in 1890. This type of rayon differs from the others by the solvent, a cupra and monium solution, used to dissolve the cellulose. This method is attributed to Swiss chemist Matthias Schweizer. The rayon, later improved upon by the J.P. Bemberg AG in 1904, made an artificial silk product most comparable to real silk fibers. Because the process contributes to water and air pollution, this rayon is not used as often as the viscose rayon, and it is no longer manufactured in the United States [3]. Cuprammonium rayon fibers differ from viscose rayons in its appearance under the microscope. The fibers, smooth and without longitudinal striations or surface grooves, are somewhat similar to cultivated silk fibers. The cross-section is round, regular, and without any visible indentations. Under the microscope, this fiber is the most similar to silk fibers. Because of its smooth, glossy surface, cuprammonium fibers polarize to a great degree, resulting in an unforgettable play of colors (Figures 4.8 and 4.9).

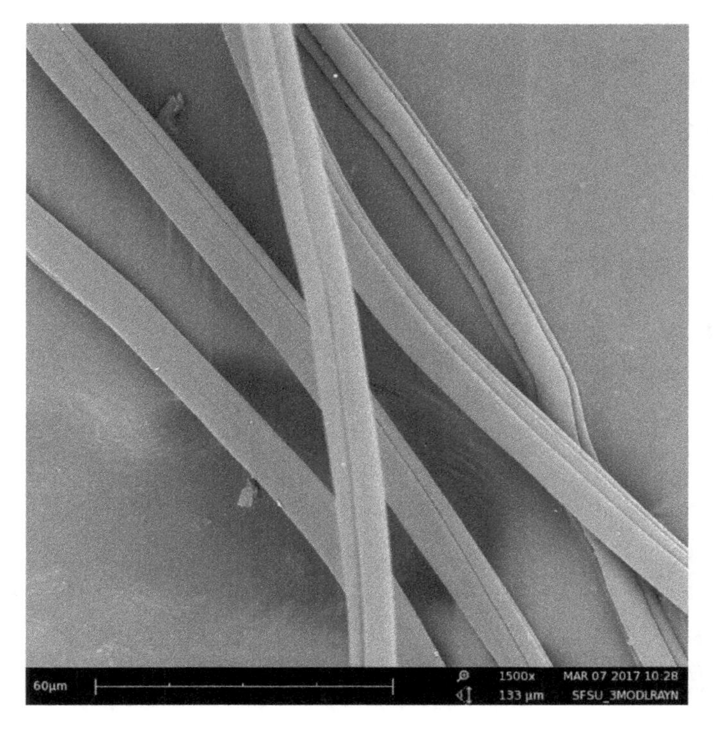

Figure 4.6 Longitudinal view of modal rayon showing fewer grooves or indentations than viscose rayon (1500×).

4.1.5 Lyocell Fibers

In the 1960s, there was a demand for a more environmentally friendly process to manufacture man-made fibers than the viscose process. The solvent used in the viscose process is carbon disulfide, a toxic chemical harmful to humans and the environment. For this reason, another method, using less toxic chemicals, was developed. To address these environmental concerns, lyocell fiber was introduced in the early 1990s. The lyocell method is *N*-methylmorpholine *N*-oxide (NMMO) based and uses monohydrate as its solvent. Under the microscope, lyocell fibers have a smooth surface and a longitudinal appearance with a round cross-section (see Figures 4.10 and 4.11). The properties of lyocell are more similar to those of cotton than any of the other regenerated cellulose fibers [3].

Ioncell-F. The viscose and lyocell processes are currently the only technologies developed for the production of manufactured cellulosic fibers. There are, however, other methods being developed, which would be superior to both

Figure 4.7 The cross-section of modal fibers has a bean or peanut shape (2000×).

viscose and lyocell methods. One example is the **Ioncell-F method** [7]. The Ioncell-F process is the first ionic liquid-based lyocell process, and it appears to be superior to the existing lyocell NMMO-based process [8]. This process is characterized by a powerful direct cellulose solvent of ionic liquids (ILs). The microscopic image of these fibers is somewhat similar to lyocell fibers as they have a round, slightly oval shaped cross-section, a smooth surface with some nodules, and small linear depressions in the longitudinal view [9].

4.1.6 A Review of Cross-sectional Shapes of Fibers

The fibers' cross-section can be engineered and manipulated when making manufactured fibers, both regenerated and synthetic. Each different shape will result in a fiber with different qualities. Because the manufacturer can manipulate the shape, these shapes can be used for a variety of manufactured and synthetic fibers, making them more difficult to differentiate when viewed under a microscope. For example, both rayon viscose and acetate/triacetate can have irregular, multilobal cross-section fibers. The cross-section's qualities

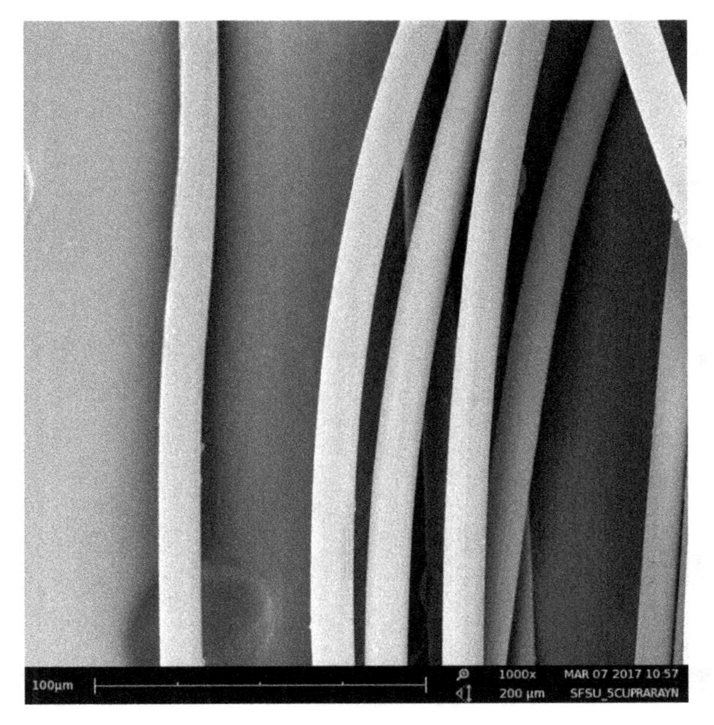

Figure 4.8 Longitudinal view of cupra rayon fibers showing smooth rod-like characteristics somewhat similar to cultivated silk fibers (1000×).

also determine the fiber's longitudinal characteristics. The irregular outline of the fibers gives them a striated appearance in longitudinal view with numerous parallel lines running along the fiber length.

The only way these similarly appearing fibers can be differentiated under the microscope is by their birefringence as rayon has much lower refractive indices (<1.500) and birefringence (≤0.005).

4.1.7 Cross-sectional Fiber Shape and Luster

Variations in the fibers' cross-sections also have an effect on how light is reflected from the fiber, ultimately determining the fabric's sheen and luster. For example, when the light hits **round** fibers, such as cuprammonium or polyester, it is unevenly reflected, which causes a **strong luster**. However, round fibers have a shiny surface because they reflect light in one general direction. Fibers with round cross-sections are rigid and not pliable, due to the regularity of the diameter of the fibers, and their smooth fiber surface results

Figure 4.9 Cross-sectional view of cupra rayon with round and sharp-edged fiber shape characteristics (1500×).

in maximum skin contact. All these qualities are aesthetically undesirable, as well as lacking in comfort (waxy, slippery to handle). For example, fiber with a **flat surface** has more luster than a fiber with a round surface. Silk has a natural **trilobal** shape, and trilobal cross-section fibers have three sides. When the light hits these fibers, they reflect it back in a direct way resulting in an appearance of high sheen. Trilobal-shaped fibers have the highest sparkling luster or sheen appearance. **Multilobal** cross-section fibers scatter the light in many different directions causing a diffuse glow with sparkles. Both trilobal and multilobal cross-section shapes result in pliable, flexible fibers, which are not stiff and soft to handle. On the other hand, a fiber with an **irregular** cross-section scatters light in many directions resulting in a dull appearance. When **Pentagonal-shaped** and hollow fibers are used in carpets, soil and dirt are less noticeable. Octagonal-shaped fibers offer glitter-free effects. Hollow fibers trap air, creating insulation and providing loft characteristics equal to, or better than, down.

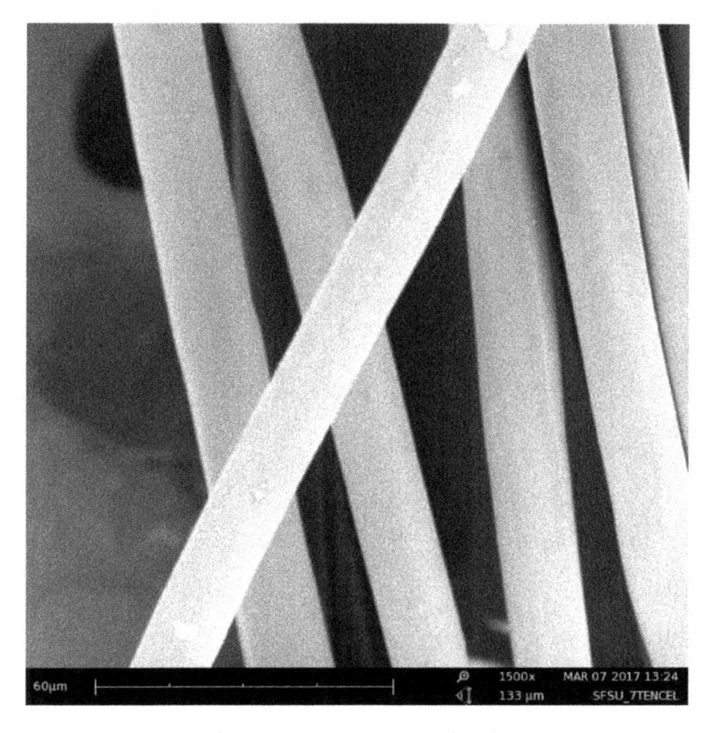

Figure 4.10 Lyocell fibers having smooth surface longitudinal appearance (1500×).

4.1.8 Acetate Fibers

Acetate fibers are the second most common manufactured fibers produced in the United States. It is the first thermoplastic heat-sensitive fiber. It was originally introduced as a kind of rayon because it is made of a cellulose derivative that starts as cellulose but is then converted into a thermoplastic substance. Note that because it is not breathable or absorbent, acetate does not share the same fabric properties as rayon, though it does have a high luster similar to silk. Because they both possess longitudinal striations, acetate fibers under the microscope may be mistaken for viscose rayon fibers. However, the striations in acetate fibers are spread further apart than those of viscose rayon. The cross-section of acetate fibers is less serrated than those of viscose rayon. It can be described as flower petal-shaped or lobular (Figure 4.12) [3]. Again, the serrated shape results from the evaporation of the solvent as the fiber solidifies in spinning. Acetate fibers are usually described as glassy and translucent filaments, flattened with few surface grooves or corrugations (Figure 4.13). The shape may vary, but when the filaments are flat, it adds glitter to fabrics [3].

Figure 4.11 Lyocell fibers with round cross-section (2000×).

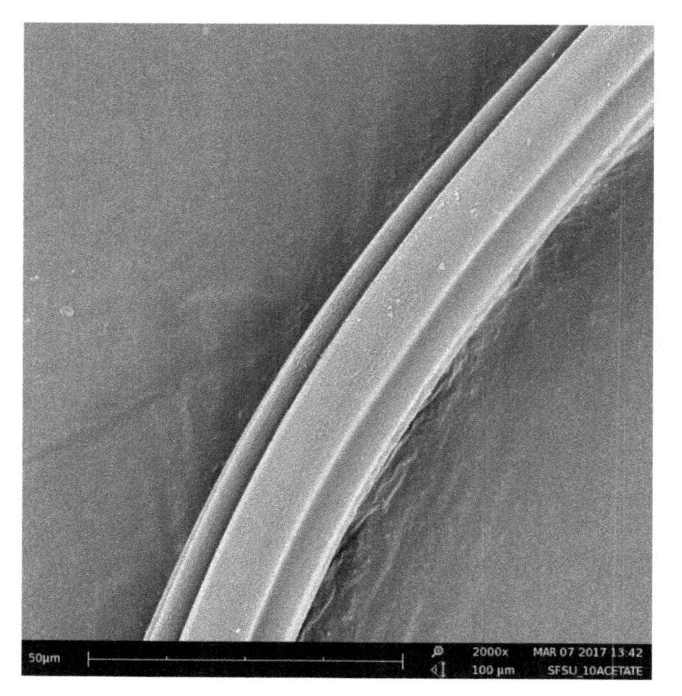

Figure 4.12 Longitudinal view of acetate fibers showing fewer striations than in viscose rayon (2000×).

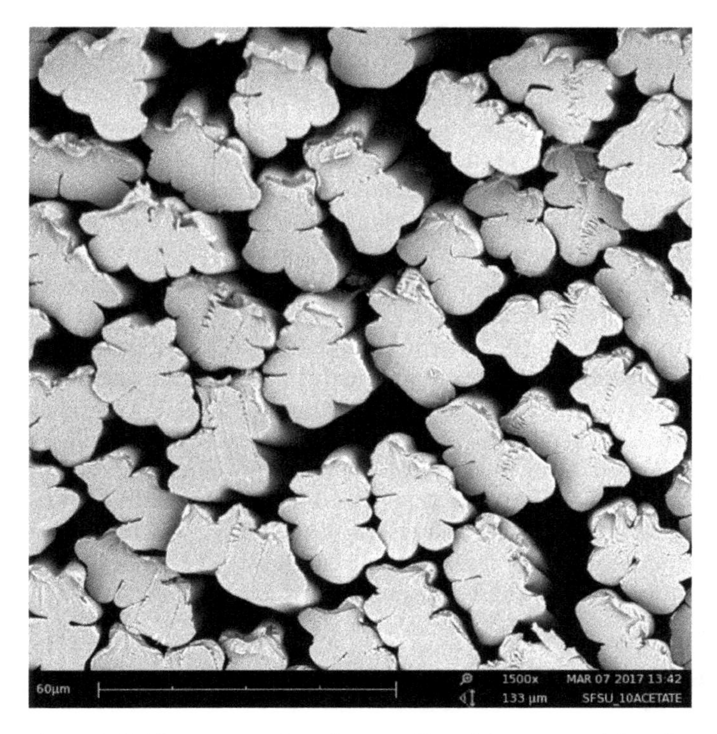

Figure 4.13 The cross-section of acetate fibers is less serrated than those of viscose rayon (1500×).

4.2 Regenerated Protein Fibers

Regenerated protein fiber, also called Azlon fiber, is defined by the Federal Trade Commission as a manufactured fiber comprised of any regenerated, naturally occurring protein. These include proteins such as casein from milk, soybean from oil manufacturing, and zein from corn starch manufacturing. These regenerated protein fibers are recognized as eco-friendly fibers because their protein substance is usually the by-products of other manufacturing processes. However, manufacturing these regenerated protein fibers requires the use of other chemical and synthetic substances. Because they are lustrous, soft, and have a good hand [10], fabrics made from protein-regenerated fibers are substituted for luxury fibers such as silk and wool. Azlon is defined by the Textile Fiber Production Identification Act as regenerated fibers derived from proteins such as milk, corn, soybean, peanut, or corn (zein). Historically, Azlon fibers were not very successful, and its production and use were discontinued. Nevertheless, they served their purpose during World War II when resources were limited. The brand name for milk fibers was Aralac, and casein fibers were substituted for wool in men's uniforms. After the war, with the increased availability of abundant resources, stronger fibers were developed ending the use of Azlon fibers, which were considered weak. Fashion historians and collectors should be aware of these fibers, especially in vintage items from around the 1930s until early 1960s. These fibers somewhat simulate wool fibers or rayons, and may be blended with other fibers. Under the microscope, these fibers have a smooth and round shape. Azlon is a generic name for regenerated protein fibers, but fashion historians and collectors, seeking to identify fibers, may find some specific brand names helpful: Aralac, Lanital, and Merinova, made from vegetable peanut protein.

Lanital (peanut protein) and other casein (milk protein) fibers appear quite similar to typical rayon fibers under the microscope, with some delustrant granules and sporadic striations. Vintage garments with Aralac and other casein fibers would usually be blended with wool. Casein (milk protein) fibers can only be distinguished from wool by microscopic examination, though chemical or burning tests will yield similar results. It is important to note that, today, there is a renewed interest in regenerated protein fibers because some are considered sustainable and environmentally friendly, even though that might not be the case. Some of these regenerated protein fibers will be discussed in Section 4.2.1.

4.2.1 Soybean Fibers

Soybean fiber is a regenerated protein fiber, derived partially from the soybean residue during the production of soybean oil and combined with a polyvinyl alcohol substance, the predominant component [11]. While soybean-regenerated fibers have high protein content, 40%, compared with other protein regenerated fibers, it is, nevertheless, predominantly a fiber manufactured by extrusion through the spinneret method. Soybean fibers were given a trademark name Soy Silk® (by the Southwest Trading Company, Phoenix, Arizona) because of the silky texture of the fibers. Under the microscope, the cross-section of soybean protein fibers is bean-shaped [10–14]. Overall, the fibers are smooth, resulting in the silky surface of fibers. However, the fibers have irregular grooves and wrinkles. These grooves will enable moisture movement along fibers. Fibers appear to have a ribbon-like characteristic, usually seen in cotton fibers, and appear to be flattened, as is consistent with the fiber's bean-like cross-sectional shape [3] (see Figures 4.14 and 4.15). They also have pronounced and elongated air-bubble inclusions. It is interesting to note that corn fibers' shape under a microscope is somewhat similar to Soy Silk fibers', especially in cross-section (see Figures 4.16 and 4.17). Soybean fibers have a soft hand and are comfortable to wear. The soybean substance gives the fibers a natural antibacterial property, similar to that of regenerated cellulosic bamboo fibers. Because soybean protein fibers have a lower moisture absorption compared with regenerated

Figure 4.14 Soybean fibers are smooth with irregular grooves and wrinkles resulting in silky surface of fibers (1500×).

Figure 4.15 Soybean fibers' cross-section is bean-shaped (1500×).

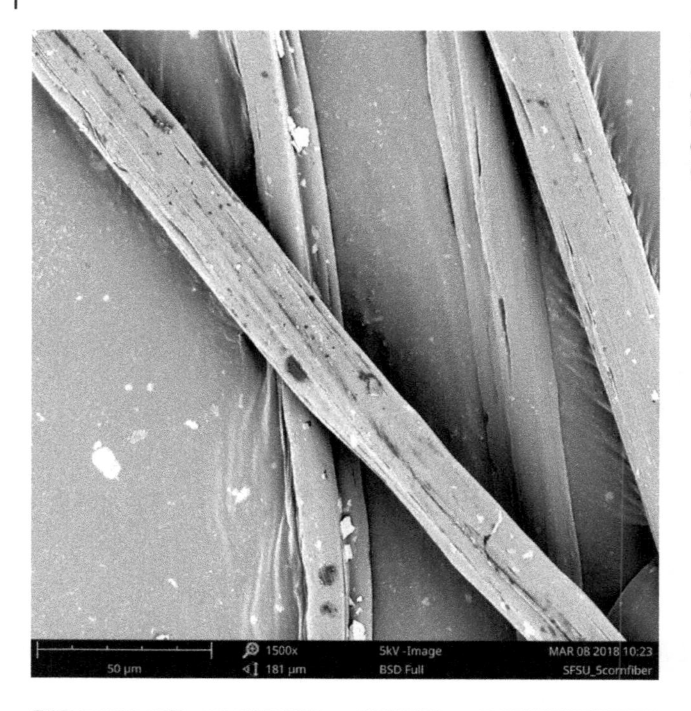

Figure 4.16 Longitudinal view of corn fibers depicting irregular groove fiber characteristics (1500×).

Figure 4.17 Corn fibers' cross-section is bean-shaped (1500×).

cellulosic fibers, they may not be as comfortable as cellulosic regenerated fibers.

A new type of rayon called **Lunacel** has been developed by Japanese scientists. This rayon is called a hybrid rayon because the fiber is both cellulosic (cross-linked cellulose cotton linter pulp) and a protein (water-soluble food protein) regenerated fiber [15]. This fiber promises a combination of the best of cellulosic and protein fiber properties. Under the microscope, the Lunacel fibers have a smooth longitudinal surface, smoother than that of silk. It is easily distinguishable from viscose rayon.

4.2.2 Milk Fibers

Milk fibers are regenerated protein fibers made from a chemical substance and casein, which is derived from milk after the butterfat is removed. The chemical substance, acrylonitrile, is also used to make acrylic fibers. Milk fibers are made using a viscose process, the same process used to produce viscose rayon. It takes 100 pounds of skim milk to make 3 pounds of milk fiber. Though some companies promoting milk fibers claim that the process is environmentally friendly, this is not the case. Because of their silky appearance and smooth hand, milk protein fibers were given the trademark name Silk Latte.

Under the microscope, milk fiber cross-sections appear similar to the bean-like shape of soybean fibers, though this shape is slightly less pronounced in milk fibers. The elongated air bubble inclusions seen in soybean fibers are not found in milk fibers. While silk fibers are translucent, smooth, or pseudo-opaque, milk protein fibers, such as soybean protein fibers, are flattened with somewhat ribbon-like characteristics [12].

In China, new milk protein fibers, from casein proteins, have been made available under a new brand name, Zhenglia. Again, as previously mentioned, the fibers are made from a copolymer of casein and acrylonitrile. These specific fibers contain a larger amount, 70–75%, of the acrylic chemical component and only 25–30% of milk proteins. However, the company still considers this process to be ecologically acceptable because no harmful chemical formaldehyde is added. The fiber under the microscope, just to add what is already described, has irregular vertical trenches and pockmarks on the surface [16]. Figure 4.18 illustrates a milk fiber with acrylic like features.

4.2.3 Composite Cellulose Fibers

Other cellulose composite fibers, called SeaCell®, have been manufactured by Zimmer AG in Frankfurt, Germany. They have been promoted as environmentally friendly fibers because they contain substances derived from algae, that is seaweed. The incorporation of seaweed provides algal ingredients that contain health-promoting effects such as anti-inflammatory and antitoxic

Figure 4.18 Longitudinal view of milk fibers with irregular vertical trenches on fiber surface (1500×).

properties. Therefore, these fibers have been utilized in the medical field and are referred to as smart textiles. The SeaCell fibers are made through the more environmentally friendly lyocell process. This is important to note because the microscopic view of the fibers is similar to those of lyocell fibers in cross-section, cross-hatchings, and the presence of fibrillation [12, 17] (see Figures 4.19 and 4.20). The cross-sections of both SeaCell and SeaCell Active fibers are identical, with a size range of 8–13 μm [12]. However, when viewing the fiber in the cross-section, the even distribution of the algae particles can be seen. These particles look similar to delustrant granules (TiO_2 pigment), which can be seen in manufactured and synthetic fibers under a microscope. The second version of the SeaCell fibers is called SeaCell Active. Silver particles are added to this second type of fiber to add an antibacterial property. This practice is very common in other smart textiles as well. The end use of a fabric with anti-bacterial property is for socks, sportswear, underwear, and home furnishings.

Other composite fibers made through a rayon viscose process are chitin cellulose fibers. Chitosan is a product derived from chitin, a natural compound derived from shellfish or crab shells. Since the chitin fibers are made through the viscose process, the cross-section looks similar to viscose rayon, with an irregular multilobed cross-section [12]. However, chitin fibers have more of a rounded shape with fewer lobes. The fiber diameters range from 10 to 15 μm.

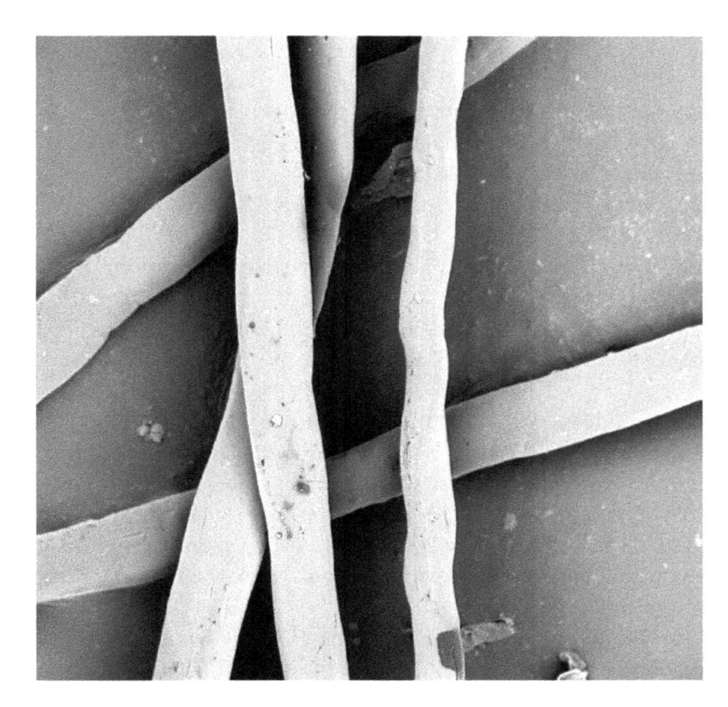

Figure 4.19 Longitudinal view of SeaCell fibers depicting smooth surface and some fibrillation with irregular rod (1500×).

Figure 4.20 SeaCell fibers with round to oval cross-section (2000×).

References

1 Xu, Y., Lu, Z., and Tang, R. (2007). Structure and thermal properties of bamboo viscose, Tencel and conventional viscose fiber. *Journal of Thermal Analysis and Calorimetry* 89 (1): 197–201.
2 Grier, W.D. (1928). Identification of rayon. *Industrial and Engineering Chemistry* 21 (2): 168–171.
3 Kadolph, S.J. and Langford, A.L. (2002). *Textiles*, 9e. Pearson Education: Upper Saddle River, NJ.
4 Erdumlu, N. and Ozipek, B. (2008). Investigation of regenerated bamboo fiber and yarn characteristics. *Fibers & Textiles in Eastern Europe* 16 (4): 43–47.
5 Afrin, T., Tsuzuki, T., and Wang, X. (2009). Bamboo fibers and their unique properties. In: *Proceedings of the Combined (NZ and AUS) Conference of the Textile Institute: In Natural Fibers in Australia*, Dunedin, New Zealand (15–17 April 2009). Textile Institute (NZ).
6 Abu-Rous, M., Ingolic, E., and Schuster, K.C. (2006). Visualisation of the fibrillar and pore morphology of cellulosic fibres applying transmission electron microscopy. *Cellulose* 13: 411–419.
7 Michud, A., Tanttu, M., Asaadi, S. et al. (2016). Ioncell-F: ionic liquid-based cellulosic textile fibers as an alternative to viscose and Lyocell. *Textile Research Journal* 86: 543–552.
8 Sixta, H., Michud, A., Hauru, L. et al. (2015). Ioncell-F: a high strength regenerated cellulose fiber. *Nordic Pulp & Paper Research Journal* 30: 30–43.
9 Hummel, M., Michud, A., Tanttu, M. et al. (2016). Ionic liquids for the production of man-made cellulosic fibers: opportunities and challenges. In: *Cellulose Chemistry and Properties: Fibers, Nanocelluloses and Advanced Materials* (ed. O.J. Rojas), 133–168. Springer International Publishing.
10 Rijavec, T. and Zupin, Z. (2011). *Recent Trends for Enhancing the Diversity and Quality of Soybean Products* (ed. D. Krezhova). Rijeka: INTECH.
11 Zhang, X., Min, B., and Kumar, S. (2003). Solution spinning and characterization of poly(vinyl alcohol)/soybean protein blend fibers. *Journal of Applied Polymer Science* 90 (3): 716.
12 Brinsko, K.M. (2010). Optical characterization of some modern eco-friendly fibers. *Journal of Forensic Sciences* 55 (4): 915–923.
13 Jiang, Y., Wang, Y., Wang, F., and Wang, S. (2004). The ultrastructure of soybean protein fiber. *Textile Asia* 35 (7): 23.
14 Vynias, D. (2011). Soybean fiber: a novel fiber in the textile industry. Agricultural and Biological Sciences. In: *Soybean – Biochemistry, Chemistry and Physiology* (ed. T.-B. Ng), 461–485. Rijeka: INTECH.

15 New rayon fiber with Lunacel (Kurabo). (2007) Chemical Fibers International, 57(6).

16 Wang, N., Ruan, C., Yu, Y. et al. (2009). Composition and structure of acrylonitrile based casein fibers. *Chemical Fibers International* 59 (2): 88–89.

17 Grieve, M. (1999). New fiber types. In: *Forensic Examination of Fibers*, 2e (ed. J. Robertson and M. Grieve), 399–419. Boca Raton, FL: CRC Press.

5

Synthetic Fibers

Synthetic fibers are petroleum-based fibers, produced entirely from chemical substances. These fibers, synthesized from petrochemicals, are made from a synthetic polymer and specifically engineered to have certain desirable properties. Synthetic fibers can be manufactured as both staple and filament. Synthetic fibers differ from regenerated manufactured fibers that are plant based such as rayon made from cellulose. Synthetic fibers are petroleum based such as polyester. Manufactured regenerated cellulosic fibers have been called first-generation fibers because they were the first manufactured fibers. Synthetic fibers have been called second-generation fibers, and microfibers have been called third-generation fibers. Microfibers are modifications of second-generation fibers (discussed in Section 5.17). The "big five" petroleum-based synthetic fibers are nylon, polyester, acrylic, spandex, and olefin (Gail Baugh [textile expert – San Francisco State University], personal communication 19 September 2017).

5.1 Nylon

The first synthetic fiber, nylon, was developed by Wallace Carothers, an American researcher who worked at the DuPont chemical company in 1930. It was a perfect time for nylon to debut as the United States was in the middle of World War II, and nylon, a fiber of high strength, could be used in military parachutes and ropes. Nylon replaced silk that was becoming scarce during wartime and was not as durable for parachute use. However, it also became popular for women's wear such as stockings. Nylon, the first true synthetic fiber, is a family of polymers called linear polyamides, derived from a diamine and a dicarboxylic acid. Nylon is chemically manufactured from organic (carbon-based) chemicals found in natural materials such as petroleum and

Textile Fiber Microscopy: A Practical Approach, First Edition. Ivana Markova.
© 2019 John Wiley & Sons Ltd. Published 2019 by John Wiley & Sons Ltd.

coal. There are two different types of nylon: nylon 6.6 (by American DuPont) and nylon 6 (by German BASF), both of which use different manufacturing methods.

5.2 Polyester

The second synthetic fiber, polyester, came about 10 years after nylon and is one of the most important developments in fiber manufacturing. Of all fibers today, 55–60% is polyester fibers. This domination of the fiber industry is due to polyester's fiber properties, such as easy care and its great versatility. One of polyester's best properties, not found in natural fibers, is that it dries fast. It is, therefore, the perfect fabric for use in athletic wear. If any other fiber is worn to the gym, such as cotton that stays wet, sweaty workout wear would not dry fast enough causing chills. Polyester fibers also have a wicking ability, moving sweat from the skin to the outer surface of the material so that moisture evaporates quickly, keeping the wearer dry. This wicking property is related to the fiber's shape, allowing manufacturers to engineer an advanced fiber shape that will increase the fibers' wicking ability. Polyester's popularity is due to its versatile nature. It is also one of the best fibers to use in blending.

While polyester was first manufactured to imitate natural fibers, its versatility has led it to become a creation in its own right. It started out as a fabric that felt like plastic, but today's polyester fabrics can feel soft and more comfortable to wear compared to earlier polyesters. One of the reasons why polyester can feel more comfortable on one's skin is due to the manipulation of the fiber shape. Fiber shape can be changed by changing the shape of the spinneret.

Polyester was introduced in 1941 by John Rex Whinfield and James Tenant Dickson, chemists in Great Britain. Its fiber-forming substance, a long-chain synthetic polymer, is composed of at least 85% by weight of an ester of a substituted aromatic carboxylic acid, including, but not restricted to, substituted terephthalic units and parasubstituted hydroxy-benzoate units (Federal Trade Commission [FTC]). Today, the most used polyester is poly-ethylene terephthalate (PET). This is also the polymer used to manufacture plastic bottles for consumer soft drinks.

It is important to understand how synthetic fibers are manufactured. Raw material (pallets) is converted into a dope or spinning solution by melting. The dope is then extruded through openings in a spinneret, a showerhead-like device with multiple holes (see Figure 5.1A). Each opening or a hole in a spinneret creates one filament, and these filaments solidify, hardening into a fiber, as they emerge from the spinneret. The shape of holes of the spinneret can be changed into any desirable shape. It can be round to create fibers with round cross-section, or it can be triangular to create fibers with triangular cross-section (see Figure 5.1B). Not only the shape of the spinneret openings/holes can be changed but also the size of the openings/holes can be changed to produce fibers of different sizes.

Because synthetic fibers, such as nylon and polyester, are chemically manufactured by extrusion through a spinneret, their fiber shapes are very uniform and similar with no distinguishing characteristics. As the spinneret determines the fiber shape, fibers can be modified to a desirable shape depending on the fiber's end use. Even though it is difficult to identify manufactured fibers through the microscopic analysis, examining the fiber shape can help us understand the fiber performance characteristics that include both aesthetic and mechanical properties. Cross-sectional fiber shape can affect not only the feel of fabric but also the performance properties of fabrics. Therefore, manufacturers change the fiber shapes according to their needs.

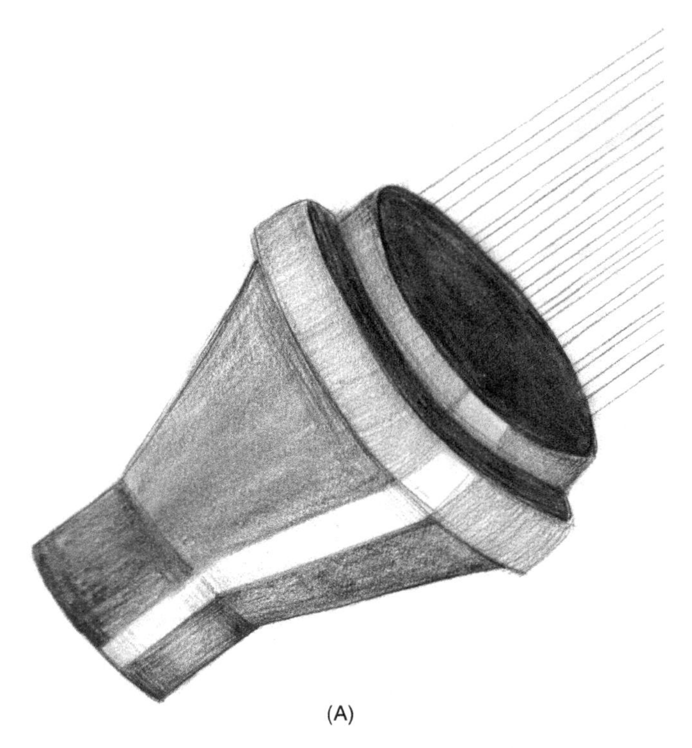

(A)

Figure 5.1 (A) Fibers extruding through a spinneret. (B) Cross-sectional view of variety of shapes in manufactured fibers ((a) hollow round; (b) triangular; (c) serrated; (d) kidney bean; (e) dog bone; (f) wavy flat; (g) square with voids; and (h) hexachannel).

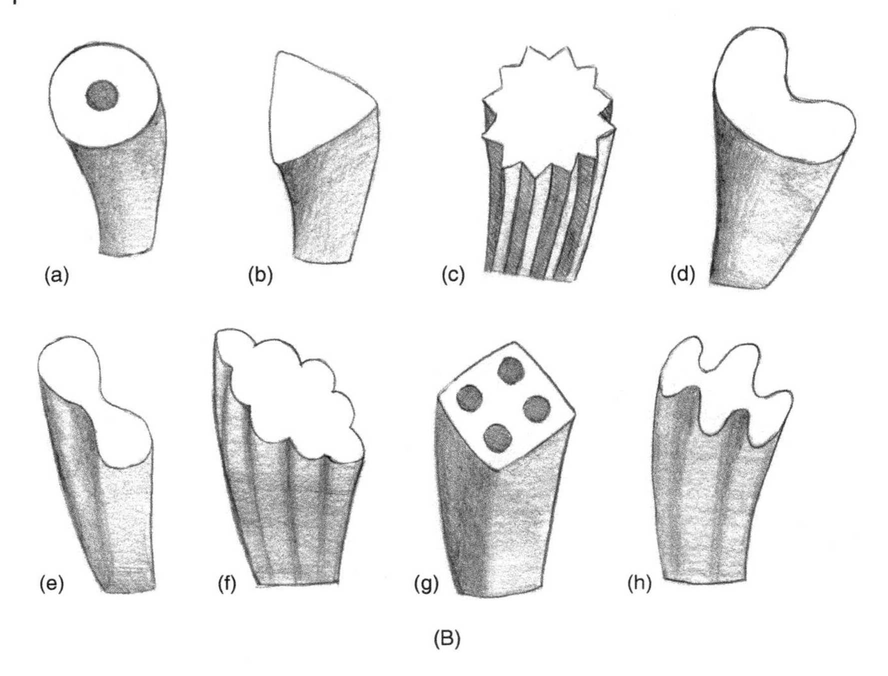

Figure 5.1 (Contd.)

5.3 Luster

Every synthetic fiber has a different degree of luster, a sheen that the fiber possesses, again depending on its fiber cross-sectional shape. The fibers that possess a high degree of luster are of round, trilobal, and oval cross-sectional shapes. The fibers' luster is proportional to the amount of light reflected by the fiber. Synthetic fibers that possess a low-degree luster are of irregular, kidney bean shaped, and octagonal shapes.

5.4 Delustering

The longitudinal view of polyester and nylon is a smooth cylindrical rod. Regular nylon fibers are transparent and under a microscope resembling a glass rod [1]. As synthetic fibers are normally bright, a delustrant is used to tone down the brightness. This delustrant, anatase titanium dioxide (TiO_2), reduces the luster in synthetic fibers. It is normally applied to the fiber spinning solution during the fiber manufacturing process before the fibers are extruded through the spinneret. Different amounts of delustrant are applied to fibers in order to vary the fiber sheen or dullness. Different luster levels are identified as bright, semidull, and dull lusters (see Figure 5.2). When viewing synthetic fibers under a microscope, the delustrant is seen as a speckled fiber surface, also referred to as delustrant granules. Under a microscope, these granules are colored dark and black. In nylon fibers, which are transparent and opaque, the delustrant will reduce this condition. While delustering helps to decrease the luster of synthetic fibers, it also weakens the fibers. Lustrous fibers without any delustrant will reflect light from its surface. Delustered fibers, with the addition of delustrant granules, will absorb the light, causing fiber degradation and tendering [1]. Thus, the tiny and uniform granules greatly reduce the wear rate.

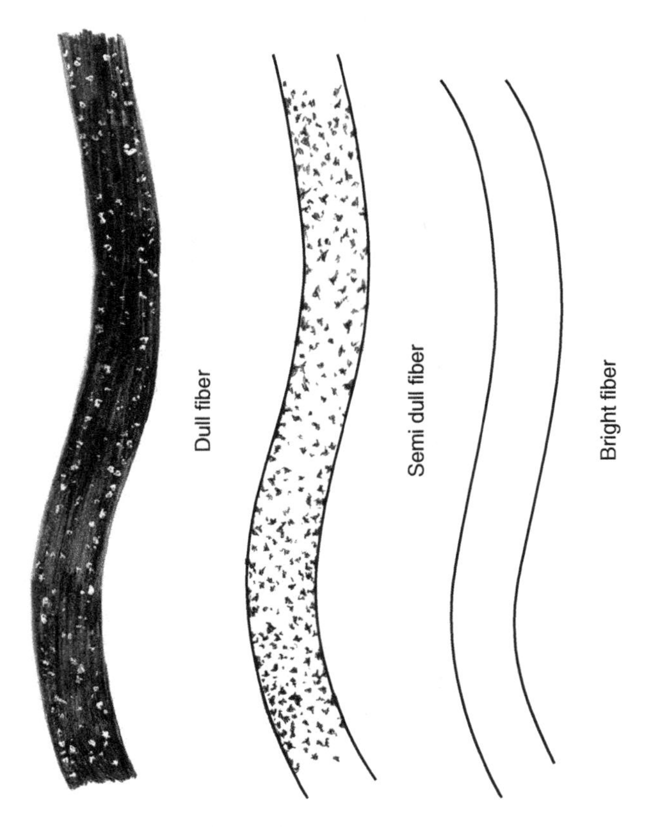

Figure 5.2 Types of fiber delustering.

5.5 Longitudinal View

The longitudinal view of polyester and nylon is a smooth cylindrical rod. Regular nylon fibers are transparent, and under a microscope, they look like a glass rod [1]. Fibers are normally bright, and to tone this brightness down, a delustrant is used. The longitudinal view of synthetic fibers usually corresponds to what the cross-sectional shape is like. For example, the most common cross-sectional fiber shape, round, will have a smooth rod-like longitudinal view. However, a trilobal cross-sectional view will produce a longitudinal fiber shape with ridges, lines, or striations running down the fiber lengthwise (see Figures 5.3 and 5.4).

5.6 Variety of Cross-sectional Shapes

What is most interesting to view under the microscope is the cross-sectional shapes of polyester and nylon fibers, although the cross-sectional shape will

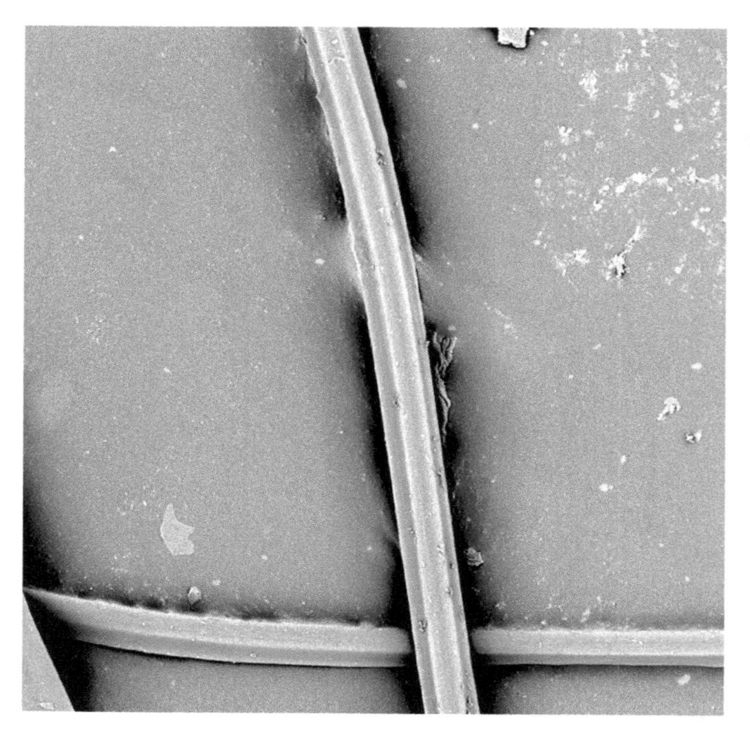

Figure 5.3 Striated longitudinal view of polyester fibers created by trilobal cross-section (1500×)

Figure 5.4 Cross-section of polyester fibers depicting trilobal cross-section (2000×)

not be an identifying factor. There is a variety of fiber cross-sectional shapes. When fiber is produced by the melt spinning method, changing the spinneret hole shape results in variations in the cross-sectional shape of the fibers. The round cross-sectional shape, the first one manufactured, created a beautiful uniform shape, which differed from what natural fibers could offer. Unfortunately, this perfect uniformity in shape created nylon fabrics with an unattractive dead feel [1]. For example, the unevenness of cotton fibers causes convolutions that give texture to cotton fabrics. However, the fabric feel can be changed by merely changing the cross-sectional shape of the fibers. Therefore, the next fiber shape that was invented was the trilobal cross-sectional shape to imitate the feel of silk fabrics (see Figures 5.5 and 5.6). Note the irregularity of the trilobal shape in natural silk fibers (Figure 5.6) when compared with the trilobal shape of the synthetic polyester fibers (Figure 5.5). The trilobal shape, used in both polyesters and nylons, adds a great amount of luster to fabrics and also adds bulk without weight, textured crimp, and increased wicking ability. Other cross-sectional fiber shapes similar to trilobal, such as triskelion or Y-shaped, possess similar characteristics [1]. The cross-sectional fiber shape can affect not only the feel of fabric but

Figure 5.5 Trilobal shape of polyester fibers – magnified (3000×) cross-section. Trilobal shape adds luster to fabrics.

also the performance properties of fabrics. Therefore, manufacturers change the fiber shapes according to their needs. To achieve good insulation in garments, such as fillers in jackets that would keep the wearer warm, a hollow fiber cross-section is needed. Hollow fibers trap air that helps retain warmth and also makes the fibers light weight with a large volume. Thermolite® polyester by DuPont is an example of insulation apparel for cold weather. It possesses high warmth and is light weight, comfortable, absorbent, and dries quickly. It is ideal for extreme weather apparel such as jackets for skiing. See Figure 5.7 for an example of hollow nylon fibers used as fillers in sleeping bags.

A flat fiber cross-section is also used in apparel synthetic fabrics. Flat fibers are formed by being pushed through a line-like hole on the spinneret. The natural fiber that this flat shape imitates is cotton fiber when their lumen is collapsed. Thus, the synthetic flat fiber imitates cotton properties such as soft garments, comfortable next to the skin. Cotton fibers move well with the natural movement of the body, their flat fiber shape naturally shaping around the body, comfortable against the skin. Many synthetic fibers also exhibit this

Figure 5.6 Trilobal shape of natural silk fibers (2000×). Note the irregularity trilobal shape in natural silk fibers compared to synthetic polyester fibers.

property, which creates fabrics with a body-skimming fluidity. The flat fiber cross-sectional shape also creates fabrics with a glimmer or glint. The very great degree of light reflected by flat fibers is very similar to how light reflects from a mirror. "Sparkling" nylon is one example of a fabric made of flat fibers [1]. Flat cross-sectional fibers have more softness, flexibility, and luster than round fibers. Figure 5.8 illustrates somewhat flat fiber cross-sectional shape of recycled/virgin blend polyester fibers.

5.7 Comparison Analysis

In comparison, fabrics made out of hollow fibers of round or trilobal shapes are not as drapable as fabrics made from full round or trilobal shapes [2]. In addition, research shows that for a better fabric feel, among four different cross-sectional shapes (round, scalloped oval, cruciform, and hexachannel) of polyester (PET) fibers, it is recommended that scalloped oval (fine fibers) fiber shape would be the most suitable for soft fabrics, and round-shaped fibers

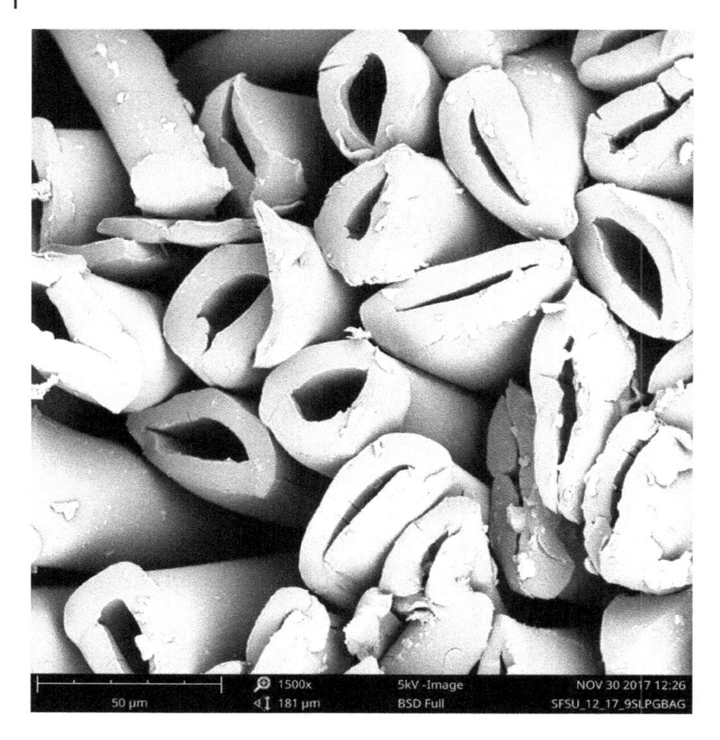

Figure 5.7 Hollow nylon fibers cross-section (1500×) – trap air and retain warmth.

(coarse fibers) would be the least suitable for soft fabrics. Cruciform shape fiber cross-section, even in its fine diameter, would not be as suitable for soft fabrics [3]. Cruciform fiber cross-section is stiffer than the ordinary round fibers. Also, hollow fibers, most specifically hollow round fibers that have higher bending rigidities than full fibers, which is why they are not drapable [2]. Again, for fabrics with high crease recovery, full round or trilobal fibers should be used [2].

Both scalloped oval and hexachannel fibers are suitable for garments that require moisture-wicking abilities. Moisture wicking allows moisture to be drawn away from the skin through capillary action and increases evaporation over a wider surface area. The hexa cross-section is a unique patterned construction with six channels that transport the moisture away from the skin and thus keep the wearer dry. As its deep-channeled structure creates a high degree of unevenness, it creates low degree of luster (see Figure 5.1B(h)). However, these are high-performance fabrics with good moisture management predominantly used in active wear. DuPont, Invista today, developed the moisture-wicking technology for polyester fabrics in the 1980s with a Coolmax®

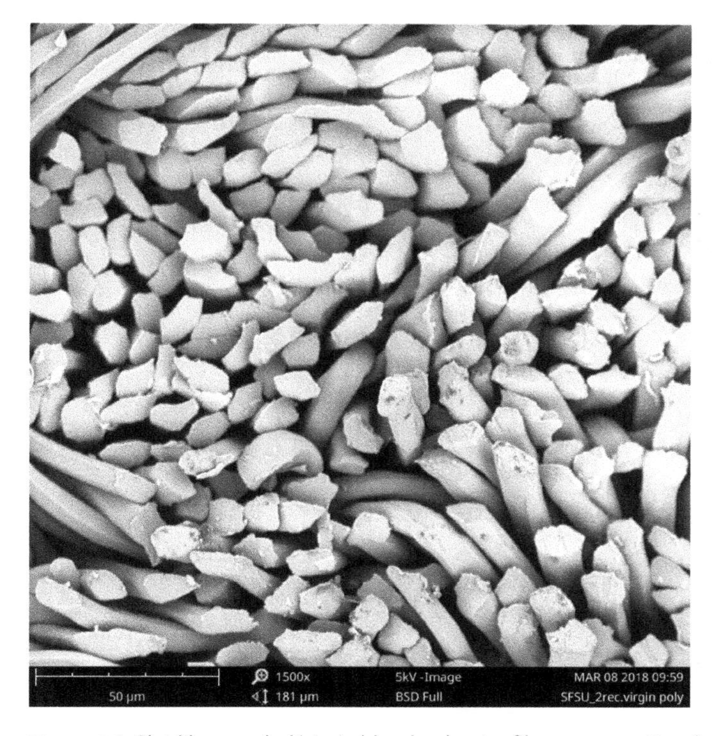

Figure 5.8 Flat-like recycled/virgin blend polyester fiber cross-sectional shape (1500×).

trademark (see Figures 5.9 and 5.10). This particular Coolmax fabric has an oblong cross-section with lengthwise grooves.

5.8 Fibers in Carpeting

The fiber shape used in carpeting is square-shaped nylon fibers with four small voids or openings (see Figure 5.1B(g)). The small voids result in good soil-hiding characteristics by holding the soil and keeping it hidden. These fibers with voids can be of square or trilobal cross-section shapes. Besides soil-hiding abilities, the voids also give the carpet thermal properties and a light weight as air gets trapped in the voids. Nylon is primarily used for carpets because of its excellent qualities such as strength and abrasion resistance. It is also important for carpets to be durable and to maintain its original appearance. Another fiber shape used very commonly in carpeting is the triskelion/trilobal shape. This trilobal cross-section in nylon fibers is by far the most popular and most efficient shape used for carpet fiber (see Figure 5.11).

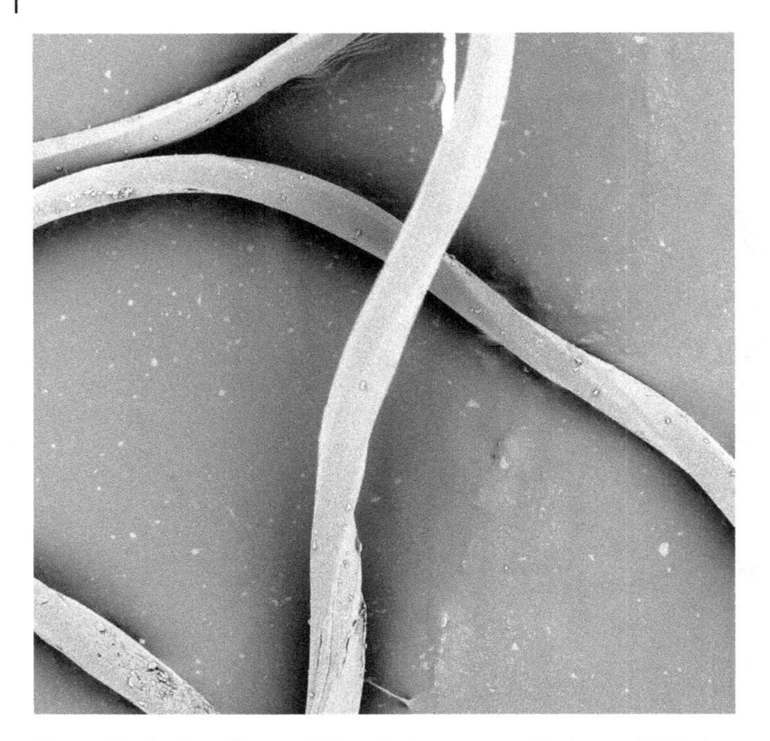

Figure 5.9 Coolmax fibers with lengthwise grooves of polyester (1000×) creates good moisture management – used in active wear.

5.9 Fabric Tenacity

The strength of fibers is also determined by the cross-sectional shape of fibers. The yarns with the highest tenacity are made from fibers with round cross-sections. Tetra and octolobal fiber cross-sections also yield yarns with high tenacity and breaking elongation [3, 4]. On the other hand, trilobal, hexsa, and cruciform cross-sectional fiber shapes make yarns with low tenacity [3, 4]. Scalloped oval cross-sectional fiber shapes produce yarns with medium tenacity [3].

When different fiber shapes are created hollow, the strength properties will also change. As we already have seen, hollow fibers trap air and are used as superior insulators. Hollow fibers are used as fillers in jackets because they keep the wearer warm. However, mechanical properties such as tearing strength and tensile strength will change when fibers are hollow. Hollow round and trilobal cross-sectional shapes have slightly higher breaking strength, while full round fibers have a better tearing strength [5].

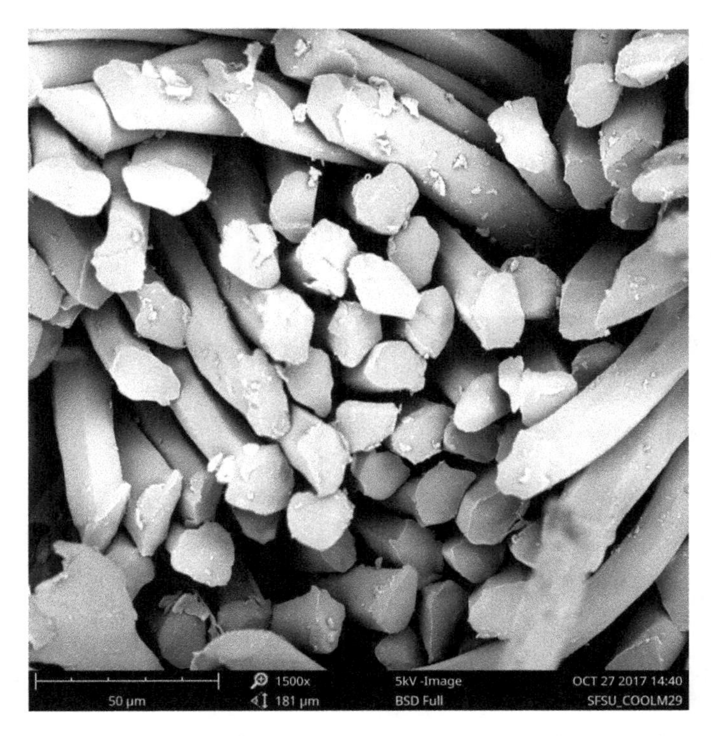

Figure 5.10 Coolmax fibers with oblong cross-section of polyester fibers (1500×).

5.10 Performance Textiles

Performance textiles are specifically engineered to fulfill a particular function, for example withstanding harsh weather conditions in icy, mountainous, and wet conditions while at the same time keeping the wearer comfortable. Teijin is a Japanese fiber manufacturer and a leader in synthetic fiber innovations. The Teijin group introduces new developments and innovations quite frequently, and therefore students are always encouraged to check their websites for updated fiber news.

The first noteworthy developments in highly modified polyester fibers is a new fiber by the brand name Octa®, whose fiber cross-sectional shape, as the name suggests, looks like an octopus. The fiber has eight "legs" that are carefully aligned in a radial pattern around the outer surface of the fiber. The core of the fiber is hollow. This unusually shaped fiber provides very unique properties. The eight legs provide rapid wicking abilities and quick drying, which is perfect for sportswear, while the fiber's hollow core also provides fabric thermal insulation and bulkiness.

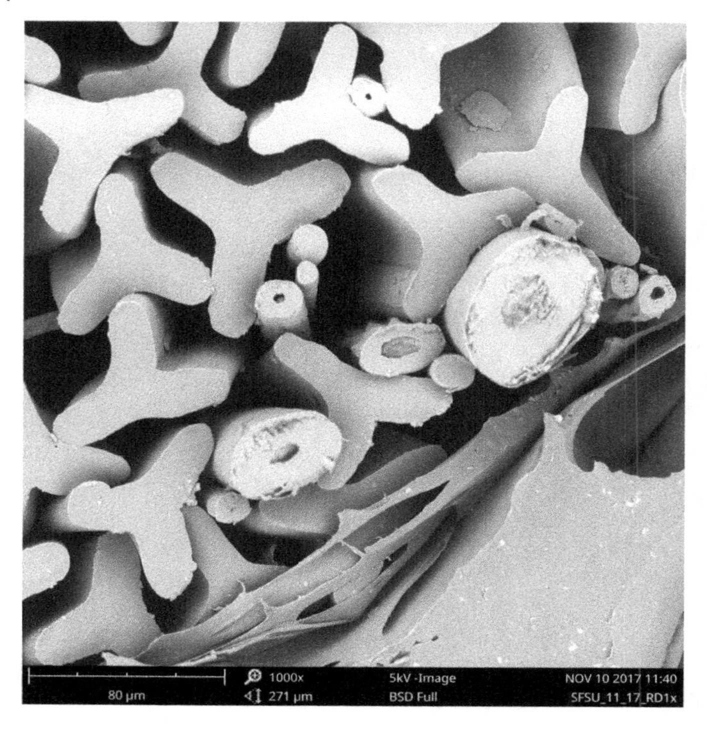

Figure 5.11 Triskelion/trilobal cross-section of nylon fibers used in carpeting. Note: Nylon fibers mixed with animal hair fibers with round cross-section (1000×).

The second notable development is Teijin's nylon fiber called WAVERON®, which again, as the name suggests, has a wave-like fiber cross-sectional shape. It is a somewhat flat fiber with four distinct waves on both sides, which increases antitransparency (see Figure 5.1B(f) for wave-like, flat fiber cross-section). Nylon fibers are inherently transparent, but this fiber is opaque, which is important in athletic wear. In addition to moisture management and quick drying abilities, this unique cross-sectional shape also adds to the softness and lightweight quality of the fabric. WAVERON is recommended for use in athleisure-type sportswear, which includes outdoor yoga wear as well.

5.11 Acrylic Fibers

Acrylic fibers were developed to imitate more expensive wool. Acrylic is known for the property of "warmth without the weight" in contrast to wool that is very heavy. Acrylic is petroleum-based synthetic fiber in which the predominant monomer is acrylonitrile (85%). Acrylonitrile is a volatile organic compound

(VOC), and an airborne carcinogen (Gail Baugh [textile expert – San Francisco State University], personal communication 19 September 2017). Acrylic is not recyclable. However, because of acrylic's low price and easy care, it is used today in a variety of consumer goods such as blankets, sweaters, socks, and sportswear (see Figures 5.12 and 5.13 for acrylic used in artificial fur).

5.12 Fiber Cross-sections

Acrylic fibers can be made in round– or kidney bean–shaped cross-section, obtained by the wet-spinning process, and also in dog bone cross-section, obtained by the dry-spinning process. The dog bone fiber cross-section gives acrylic fibers the bulkiness that is needed to imitate wool's bulkiness (see Figure 5.14 for a dog bone cross-section in an acrylic blanket). The dog bone cross-sectional shape allows fibers to avoid packing closely. This lower packing ability creates a bulkier fabric. A lower fiber-packing ability is suitable for garments worn during high-stress activities in which much sweating occurs

Figure 5.12 Longitudinal view of acrylic fibers (used as artificial fur) depicting texture on fiber surface (2000×).

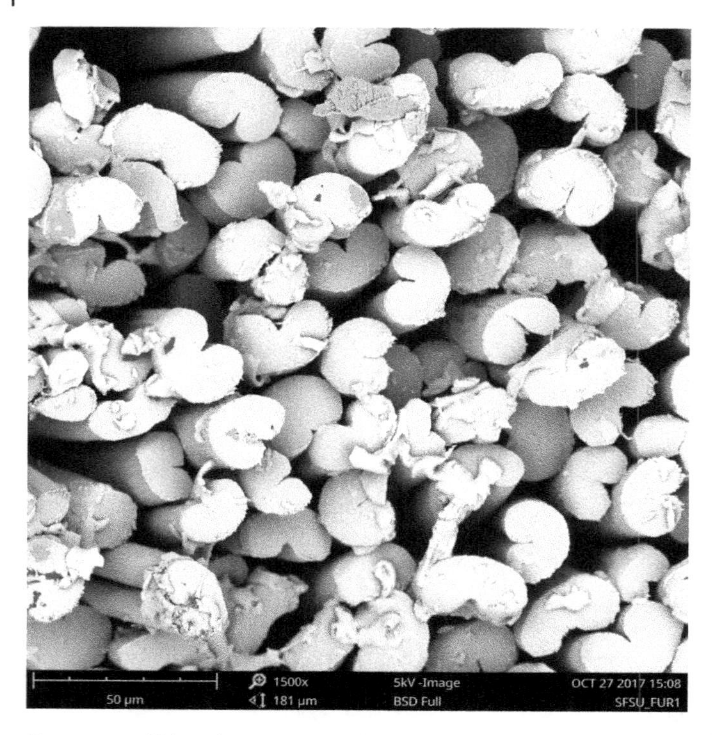

Figure 5.13 Kidney bean cross-section in acrylic fibers (used as artificial fur) (1500×).

and large amounts of sweat need to be removed from the skin for the wearer to be comfortable. As they do not pack closely, the large spaces in between the bone-shaped fibers allow for larger amounts of perspiration to wick away [6]. The dog bone's narrow cross-sectional fiber shape results in exceptional softness with a glimmer, lending fabrics a desirable softness and luster. The kidney bean–shaped cross-section also creates bulk in acrylic fibers, making them suitable for blankets and bulky sweaters. They also possess a higher bending stiffness that increases fiber resiliency [1]. The kidney bean–shaped cross-section is the most similar to cotton's cross-section of fibers and therefore will add the next-to-the skin, comfortable feel to the fabric (see Figure 5.13 for kidney bean cross-section shape in acrylic fabrics).

5.13 Fiber Longitudinal View

Acrylic fibers were manufactured to imitate wool. As wool has a texturous crimp, in hand and in appearance, acrylic fibers also needed to be texturized. Synthetic fibers have to be further modified after they come out of the

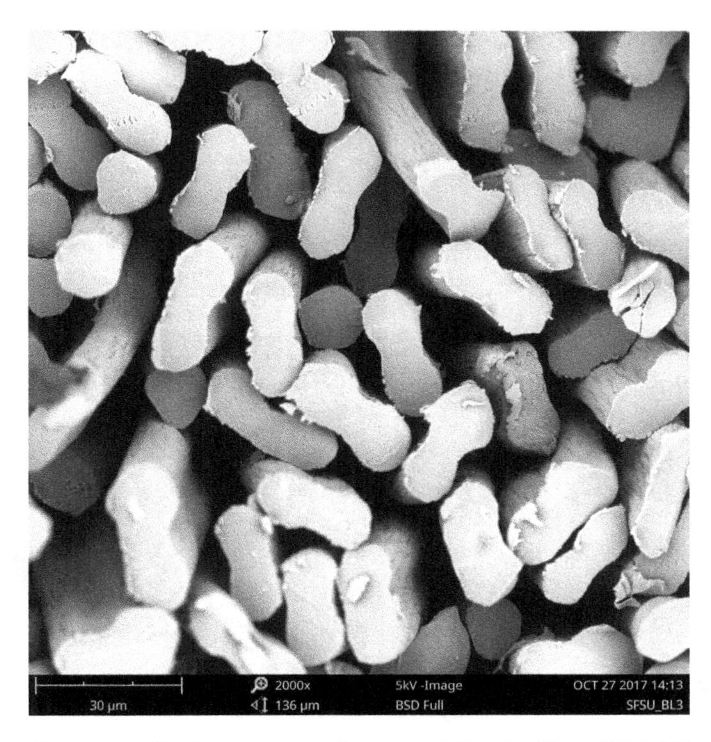

Figure 5.14 Dog bone cross-section in acrylic blanket fibers (2000×). Dog bone cross-section shape adds bulkiness to imitate wool's bulkiness.

spinneret, as they normally come out straight and smooth. This can be done by adding heat to the fibers, which will bend the fibers creating texturous surface. A texturous or fuzzy fiber surface can be seen under an optical microscope (40×). Rather than having a smooth surface, as we would see in nylons and polyesters, acrylic fibers have a texturous surface when viewed longitudinally under a microscope (see Figures 5.12, 5.15–5.17). When viewing acrylic fibers under a higher magnification such as 3000×, small cracks are visible on the fiber's textured surface. As these cracks widen, tiny portions of the fiber break off under abrasion causing acrylic fibers to pill.

Modacrylic fibers are modified acrylics. They are made of acrylonitriles, from 50% to 85% acrylics, with varying amounts of other polymers added to create the copolymer. They are the first inherently flame-retardant synthetic fibers and therefore the most suitable for flame-resistant interior design goods such as window treatments. Modacrylics can also create fur-like fabrics and are extremely useful in wigs and faux-fur garments. Today, thanks to modacrylic fibers, faux-fur has a very realistic appearance. Separate guard hairs and fur hairs (under hair) can be created, closely resembling real furs. The texturizing

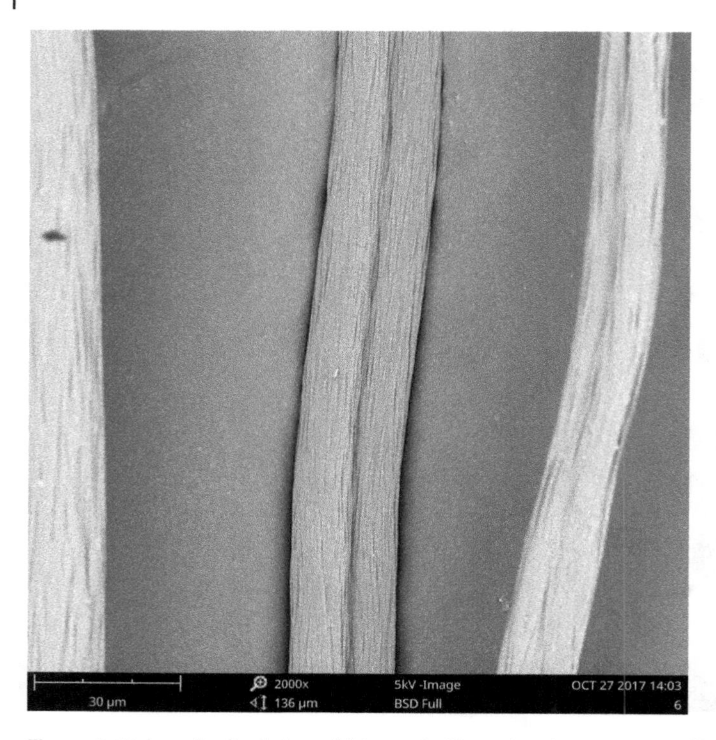

Figure 5.15 Longitudinal view of felt acrylic fibers showing texture on fiber surface similar to fibrous surface, which can crack and fiber may break off and pill with wear (2000×).

of fibers also occurs with modacrylics to create fur-like and hair-like fibers. Under a microscope, modacrylic fibers have a dog bone, irregular/serrated, or U cross-section fiber shape. Because modacrylics are creamy white, if a delustrant is added, dark specks would be seen in cross-section and in longitudinal views under a microscope. Because some fibers can be slightly serrated in cross-section, the fiber surface in longitudinal view might appear striated or grainy.

Crimp is created in a variety of ways, including manufacturing bicomponent fibers. **Bicomponent** acrylic fiber is defined as a fiber made of two different polymers. These polymers differ in either physical or chemical properties or in both physical and chemical properties. The two polymers, extruded from one spinneret, are side-by-side, making one filament. In summary, bicomponent acrylic fibers are manufactured by combining two different acrylic polymers, extruded side-by-side to form one fiber. The two acrylic polymers will have different heat resistance with one melting faster than the other. This is how they create crimp, varying from deep crimp to lesser crimp.

Figure 5.16 Cross-sectional view of acrylic fibers – felt imitation, depicting lobed cross-section SEM 2000×.

5.14 Spandex

Spandex was developed because of the need for synthetic replacements for rubber. Spandex can be made into smaller deniers than rubber. Known as elastane all over the world, in North America it is referred to as Spandex. DuPont Industries introduced a variation of Spandex by the trade name Lycra in 1958. Elasticity and elastic recovery are the most important properties of Spandex that can be stretched 500% and recover without breaking. Spandex is used in apparel products that require elasticity such as swimwear and active wear and also in fashionable apparel such as leggings, skinny jeans, and tight-fitting dresses. Spandex fibers have a wicking ability and are comfortable to wear. It is a good candidate for use in lingerie. Spandex fibers are used in combination with other fibers, which are manufactured in a variety of ways. As the cost of spandex is high, the bare core spandex makes up just a small percentage of the fiber content of a garment.

Figure 5.17 Longitudinal view of acrylic blanket fibers depicting (typical in acrylic) texture on fiber surface (2500×).

Combinations of Spandex with other fibers can be made in monofilament or in multifilament form. Monofilament Spandex fibers have a round cross-sectional shape. Multifilaments are fibers that are fused together. Under a microscope, they resemble a thick fiber with many striations along the fiber, whereas a multifilament Spandex fiber resembles fused fibers bunched together (see Figures 5.18 and 5.19 for cross-sectional views of Spandex fibers). There are three fused round Spandex fibers in both figures. In this fabric, Spandex fibers were blended with nylon fibers that have a trilobal cross-section. Notice the large fiber diameter of the Spandex fibers when compared with the nylon fibers. The number of filaments that can be fused together to create the multi-filament fiber can vary from a low of 12 fibers to a high of 50 fibers. Usually, spandex fibers are covered with a different yarn and Spandex as the core, as is also done with a double yarn. Spandex fibers have round cross-section. Spandex fibers are usually of white color and have dull luster.

Figure 5.18 Cross-sectional view of spandex/nylon blend. Spandex fibers are of round cross-section and are fused depicting a bunched of three fibers together. Smaller in diameter – nylon showing trilobal cross-section (1500×).

5.15 Olefin

Olefin (polypropylene) fibers are the least used in textiles of the big five synthetic fibers. Olefin is the least expensive of these synthetic fibers as it is the least complex to make compared with other synthetic fibers. However, its durability is not as good. For example, in contrast with higher-durability nylon, a carpet made from olefin fibers would be cheaper to buy but would not last as long as a carpet made from nylon fibers.

Similar to other thermoplastic fibers, olefin fibers are hydrophobic. However, they also have a waxy surface so that materials made out of them can be easily wiped clean with a cloth. Olefin fabrics are found in technical uses such as envelopes and shower curtains. Because of their hydrophobic nature, olefin fibers tend to be quick drying and have an instant wicking ability. Therefore, in apparel products, olefin fibers are used in active wear, warm clothing, socks,

Figure 5.19 Cross-sectional view of spandex/nylon blend. Spandex fibers are of round cross-section and are fused sideways depicting a caterpillar–like cross-section. Smaller in diameter – nylon showing trilobal cross-section (2000×).

and linings. Because of its great colorfastness, olefin is also suitable for interior design applications. However, due to its crystalline structure, the fibers have poor dyeability and must be colored in fiber or solution stage. Because colors can be applied only in the earliest stages of fiber production, colors must be chosen ahead of the fashion trends with limited color flexibility. Olefin is also manufactured in monofilament and multifilament fiber forms. Monofilament olefin mainly has a round cross-sectional shape (see Figures 5.20 and 5.21). In longitudinal view, the fiber is smooth with a waxy feel. Olefin fibers can also have a triangular cross-section when used as outdoor fabrics for interior design products.

5.16 Fiber Melting Point

As previously mentioned, synthetic fibers cannot be distinguished under a microscope because the nozzle opening of the spinneret can be of any desirable

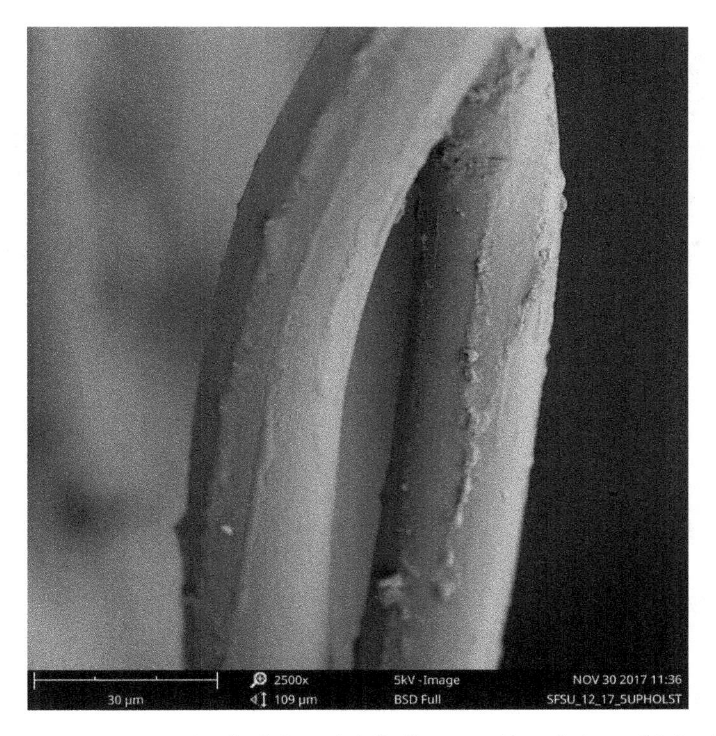

Figure 5.20 Longitudinal view of olefin fibers used in upholstery fabrics (2500).

shape. Thus, synthetic fibers end up having similar physical characteristics under a microscope. For example, both nylon and polyester can be made of round and trilobal cross-sections. For this reason, other identification methods, such as the hot plate method, are used for synthetic fibers.

As thermoplastic fibers melt under different temperatures, their melting point can be used as a distinguishing factor when identifying thermoplastic fibers. This procedure is somewhat difficult because fibers do not melt at one temperature, but they melt over a temperature range. The higher the degree of fiber crystallization, the higher the melting point. To observe the melting points of thermoplastic fibers, a stereo microscope and a stage slide warmer is necessary. Fiber samples should be cut into a small size of around 2 mm. A mounting medium of a high-temperature stable silicon oil is recommended. When observing the melting process under a stereo microscope, a melting range is defined as from when the melting begins until no further changes occur. The examiner might see other thermal reactions such as softening, bubbling, and charring during the melting process. The test ought to be repeated around 10 times to obtain values as accurate as possible. Melting point determination is quite accurate for distinguishing between nylon 6 and nylon 6.6

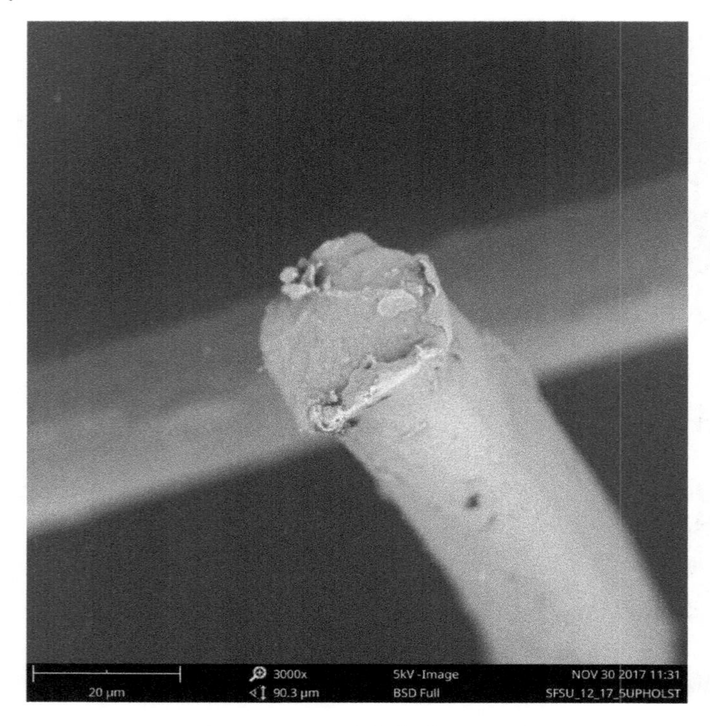

Figure 5.21 Round cross-section of Olefin fibers used in upholstery fabrics (3000×).

because nylon 6.6 has a significantly higher melting point. However, the melting point determination becomes somewhat inaccurate when identifying nylon 6.6 and polyester because their melting points are very close. Acrylic fibers do possess some distinguishing characteristics such as the texturous surface in their longitudinal view, which could be used as fiber identification points. Because acrylonitrile, of which acrylic is composed, is considered a carcinogen and is airborne when heated, for health reasons, it is not recommended that acrylic fibers be burned or melted in a laboratory. See Table 5.1 for melting temperatures.

5.17 Microfibers

Microfibers (MFs), developed in the 1990s, were one of the greatest achievements of the manufactured fibers' industry. The introduction of microfibers started a new era of textile products. Microfibers are thinner than any other natural or manufactured fiber. While the first microfiber was made of

Table 5.1 Melting points of various synthetic fibers.

Fiber type	Melting temperature range	
	°F	°C
Nylon 6	410–428	210–220
Nylon 6.6	490–510	255–265
Polyester	482–550	250–288
Acrylic	—	—
Spandex	446–518	230–270
Olefin	320–350	160–177

polyester, today other fibers can also be produced in microfiber form, including nylon, acrylic, and some regenerated cellulosic fibers such as rayon and lyocell (see Figure 5.22 for polyester microfiber 5000× magnification). Up until the development of microfibers, the finest fiber was silk (see Figure 5.23 for relative fiber size comparison, and see Figure 5.24 for microscopic view of

Figure 5.22 Longitudinal view of magnified (5000×) polyester microfiber.

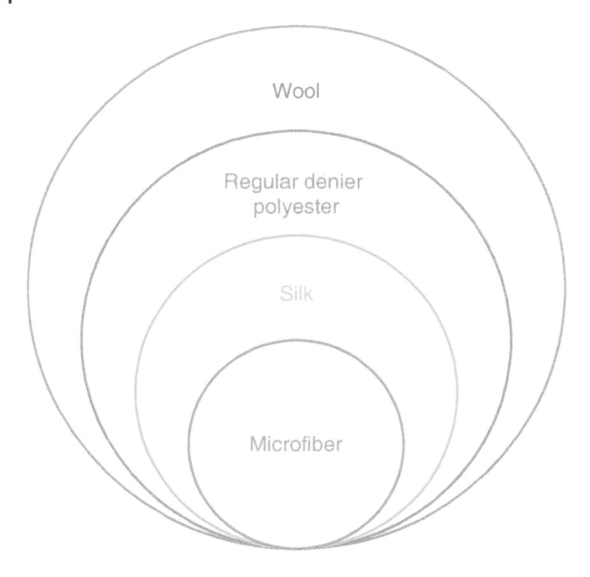

Figure 5.23 Relative size of microfibers compared with other fibers (approximation of fiber diameter).

microfibers and regular denier fibers). Microfibers, characterized by the fiber diameter and also referred to as microdeniers, are finer than silk and are defined as fibers that range in size from 0.1 to 1.0 dtex. Microfibers smaller than 0.1 are regarded as ultra microfibers (UMFs). Microfibers can be viewed under a simple light microscope with a magnification of 40× or higher. Again, microfibers' fiber content cannot be identified via microscope because, as is the case of manufactured regular denier fibers, they are extruded from a spinneret with the spinneret's shape determining the fiber's shape. However, it is still important to know how the shape and smaller size of the fiber affects the fiber's properties and also the fabric's end use.

Even though the first microfibers were invented in the 1960s by Miyoshi Okamoto, a scientist, in the Toray Industries in Japan [7], the fibers were not used by the industry until the 1990s. The first microfiber was produced by Hoechst A.G. of Germany in mid 1980s, followed by DuPont in the United States in the early 1990s [8]. The success of microfibers in the textile industry is due to their properties, surpassing those of natural fibers. Microfibers' extra softness, luster, pleasant hand, and good draping qualities make them suitable for blouses, dresses, undergarments, and bed sheets. Microfibers not only surpass the properties of natural fibers, such as softness and good drape, but they also surpass the properties of other regular denier manufactured fibers, such as dimensional stability and durability. Microfibers also have an increased wicking ability and performance characteristics, such as strength, compared to regular denier fibers, making them ideal for performance textiles such as sportswear.

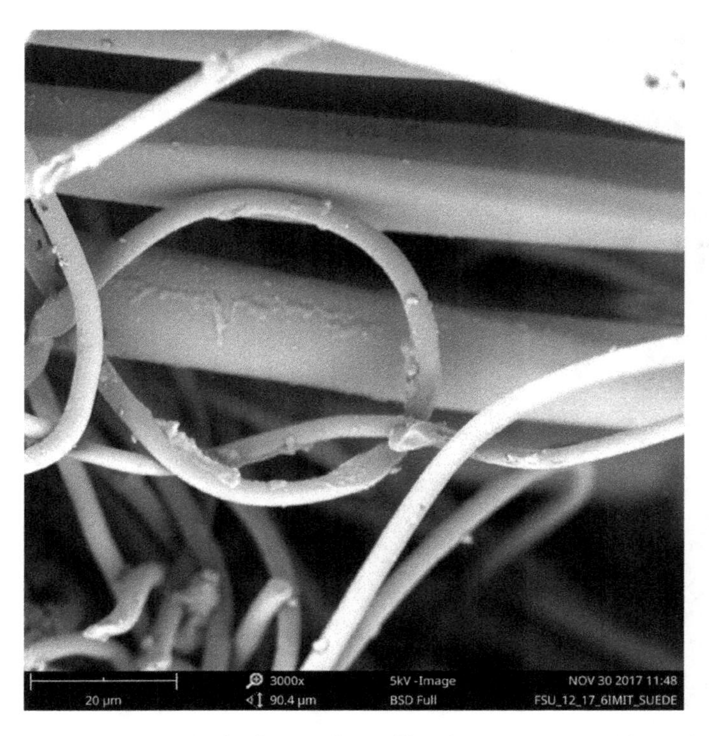

Figure 5.24 Longitudinal view of microfiber size comparison with regular-denier fibers (3000×).

Microfibers allow synthetic fabrics, such as nylon, to have a feel and soft hand comparable to cotton. One such a fabric is a high-performance fabric, known by the trade name Supplex® by Invista, made for active wear. This fabric is made out of nylon microfiber and is known for having a cotton-like hand (see Figures 5.25 and 5.26).

TACTEL®, another example, is a microfiber having noncircular cross-sections. It is an adapted form of nylon 6.6, developed for apparel rather than for use in carpets or upholstery. TACTEL is a registered trademark of Invista for soft, lightweight nylon fibers (see Figures 5.27 and 5.28). The cross-sectional fiber shape is more irregular, which adds texture to the fabric. Because of its soft hand, TACTEL is used in the intimates' product category, specifically in undergarments. As it dries faster than cotton, it is useful as a replacement for cotton. It possesses the softness of cotton and the strength, durability, and easy care of nylon. It is not only the fiber cross-section that modifies the performance characteristics of this nylon but also the fact that it is a microfiber and not a regular denier fiber; microfibers produce softer fabrics.

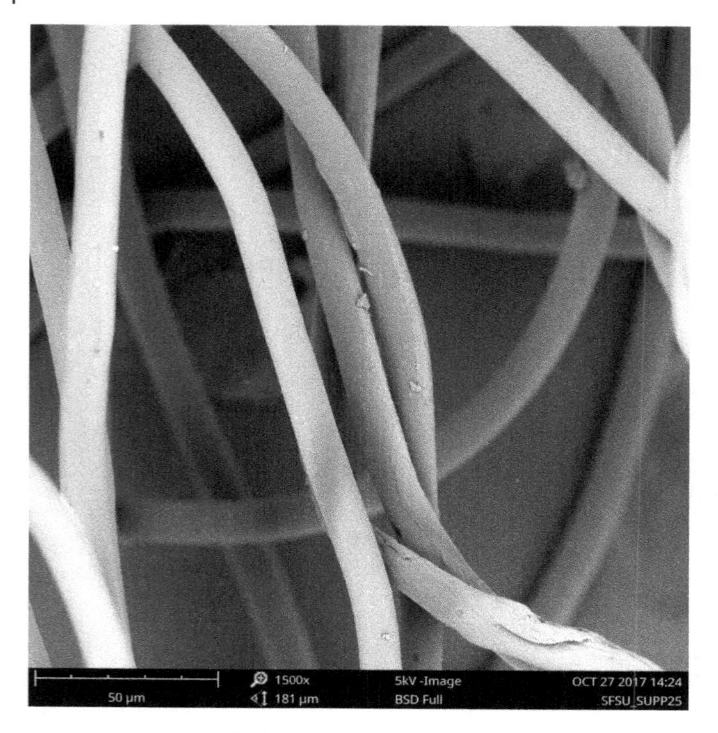

Figure 5.25 Longitudinal view of nylon microfiber Supplex – its small diameter allows fabrics to achieve a cotton-like hand fabrics (1500×).

5.17.1 Applications of Microfibers

Microfibers did not only make their debut in sportswear for their wicking abilities but also became a great candidate for rainwear. When microfibers are tightly woven, they prevent water droplets from penetrating the fabric weave. However, there are still openings small enough for water vapor to infiltrate through the weave. This exceptional water impermeability property in microfiber fabrics makes them a suitable fabric for waterproof wind breakers. Microfibers allow to make fabrics that are both water- and wind-resistant. Microfibers also provide a great amount of surface area because there are more microfibers in a yarn than there are regular denier fibers in a yarn. This property makes microfibers also suitable for insulation [9].

Microfibers are produced in filament and in staple forms. The creation of **filament** microfibers was prompted by the possibility of creating a synthetic fabric that is as close to natural silk fabric as possible. These attempts, which started in Japan, resulted in microfibers called Shingosen, which includes fabrics that are highly aesthetic and possess sensual factors with a superior touch,

Figure 5.26 Nylon microfiber Supplex showing irregular cross-sectional fiber shape (2000×).

handle, texture, and softness [10]. These are the new generation microfibers with noncircular cross-section, made of splitting a bicomponent filament. The Shingosen trend began in Japan, but later spread worldwide [10]. **Staple** microfibers have also been explored not only to as a substitute for silk but also as substitutes for other natural staple fibers such as cotton, wool, and flax [9].

Microfibers are used alone, or they are blended with other regular denier fibers. Figure 5.24 illustrates the size differences in microfibers and regular denier fibers. Microfibers can be blended not only with other synthetic fibers but also with natural fibers such as wool. Blending polyester staple microfibers with wool, fine wool suiting for example, is not a mere blend that makes the wool fabric cheaper; it actually results in a wool fabric having a finer hand [9]. Blending microfibers with regular denier fibers adds softness and good drape to the fabric.

The technology for manufacturing microfibers is more sophisticated because the fibers themselves are more delicate. Microfibers are manufactured by two main processes. The first one, often used for polyester, nylon, and acrylic microfibers, is simply the conventional melt spinning method, extruding fine filaments from a spinneret with very small holes of a fine diameter and then

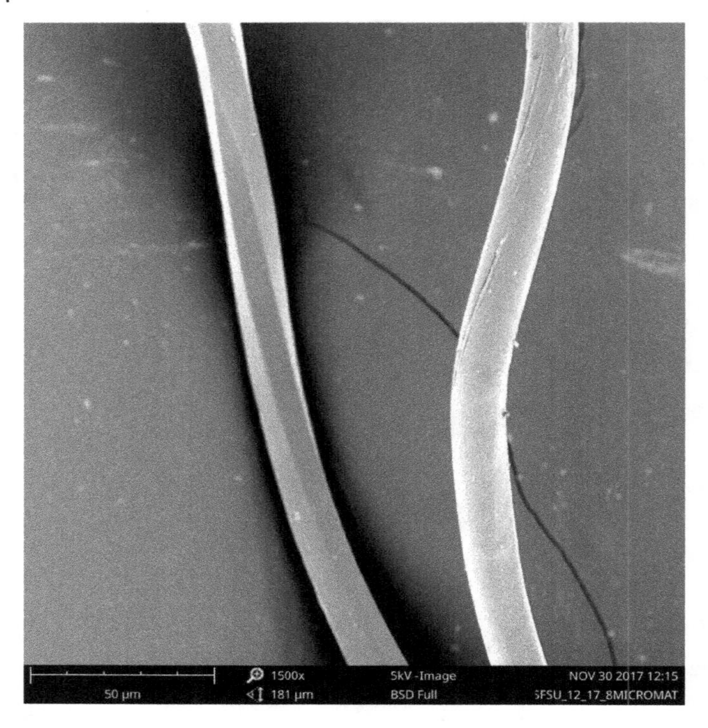

Figure 5.27 Microfiber on the left with irregular cross-section (TACTEL nylon), and microfiber on the right with round cross-section (Micromattique polyester) (1500×).

further drawing them at a high ratio to make them finer. In order to withstand this method, the polymer must be of high quality.

The second process, the split method, creates fibers with two different polymers. After the fabric is made through a finishing process, the fibers are split into much finer fibers. A fiber made out of two different polymers is called a bicomponent fiber. The advantage of using a bicomponent fiber enables the fiber and also the fabric to contain properties of both materials. Bicomponent fibers offer modified properties while maintaining the major polymer properties.

Bicomponent fiber is defined as a fiber made of two different polymers that have either different physical or chemical or both physical and chemical properties, for example polyester and nylon. The two polymers are extruded from one spinneret to make one filament. However, bicomponent fibers may also be made of two variants of the same generic fiber such as acrylic bicomponent fiber that creates permanent crimp. These two polymers are extruded from a specially engineered spinneret of a variety of configurations. These two

Figure 5.28 Cross-sectional view of irregular shape microfibers (TACTEL nylon) blended with round-shaped microfibers (Micromattique polyester) (2500×).

polymers can be of different configurations, for example side-by-side, core-sheath, or multilayer.

The **side-by-side** bicomponent configuration is composed of two fibers lying side-by-side. In the case of acrylic, a bicomponent fiber that creates permanent crimp, this is achieved through the side-by-side bicomponent technique. Two different acrylic types are used, which have a different heat resistance where one melts faster than the other and thus creating crimp in the fiber. Figure 5.29 illustrates the examples of bicomponent fibers.

The **Core-sheath** bicomponent configuration is one where the polymers are the core and the second polymer is the sheath, completely surrounding the core. The configuration of polymers can vary in the end use of the fiber. For example the core can include a less-expensive polymer to bring down the fiber cost, and the sheath can be made from an expensive polymer making the fiber visually appealing. Or in another example, the core is made of polyester, which imparts strength, and the sheath is made of PET–PEG (polyethylene glycol) polymer, which contains antistatic and antisoil abilities.

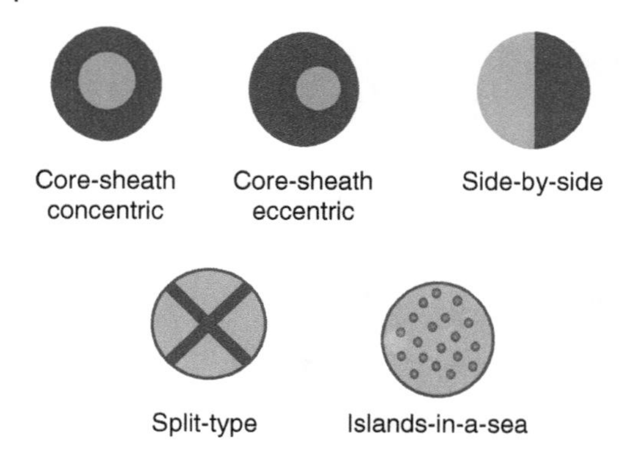

Figure 5.29 Cross-section of several types of bicomponent fibers

Core-sheath can be created in a variety of geometries in addition to concentric circles geometry depending on the end-use application. Another one would be eccentric geometry. If the product goal is to maintain strength, concentric geometry is applied, whereas if the product goal is to add bulkiness to the end product, the eccentric geometry is applied to the bicomponent fiber.

Some bicomponent fibers are made from a combination of polyester and nylon. They each add their properties to the final product. For example, microfiber cleaning cloth has been created, as a polyester and nylon bicomponent fiber. The polyester adds the oil absorbent property, and nylon adds some (10%) water absorbent property. Even though nylon is a hydrophobic fiber, of all synthetic fibers, it would absorb the most water in microfiber cleaning cloth.

Another interesting type of indirect spinning is **island-in-a-sea** type that is composed of the island component and the sea component. The island polymer is usually polyester or nylon and the sea is of polyvinyl, polyolefin, or polyester copolymer [7] content. In this process after the filament is extruded from the spinneret, the sea is consequently dissolved, and only the islands stay behind, which are basically fine microfibers (see Figures 5.30 and 5.31). These microfibers can range in size from 0.1 to 0.01 dtex.

The **split-type** microfibers also involve a more complicated processing method. In this process, two different incompatible polymers are used, and after they are extruded through the spinneret, they are separated or split into smaller fibers. This split-type fiber has flower-like shape, which consists of two parts – the star-shaped core, made of nylon, and the wedge-shaped parts, the flower petals made of polyester. These two parts are split at some point of time after it is extruded from the spinneret (see Figure 5.32).

Cleaning cloth is known for using this split type of microfiber because small dust particles get picked up in between the flower petal wedges of the

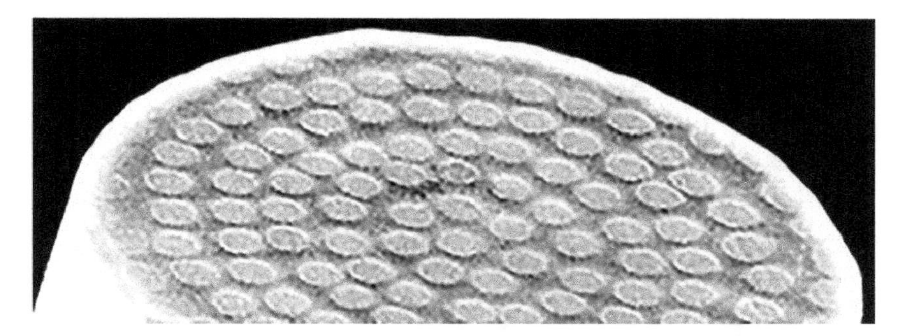

Figure 5.30 Cross-sectional shape of island-in-a-sea type microfiber prior to being dissolved. *Source:* Courtesy of Toray Industries.

Figure 5.31 Longitudinal view of island-in-a-sea type microfiber after being dissolved – only the islands stay behind. *Source:* Courtesy of Toray Industries.

microfiber. The properties of both polyester and nylon are utilized in this bicomponent fiber as polyester absorbs oil stain, and nylon can absorb some water particles (10%) even though nylon is a hydrophobic fiber. This is how the

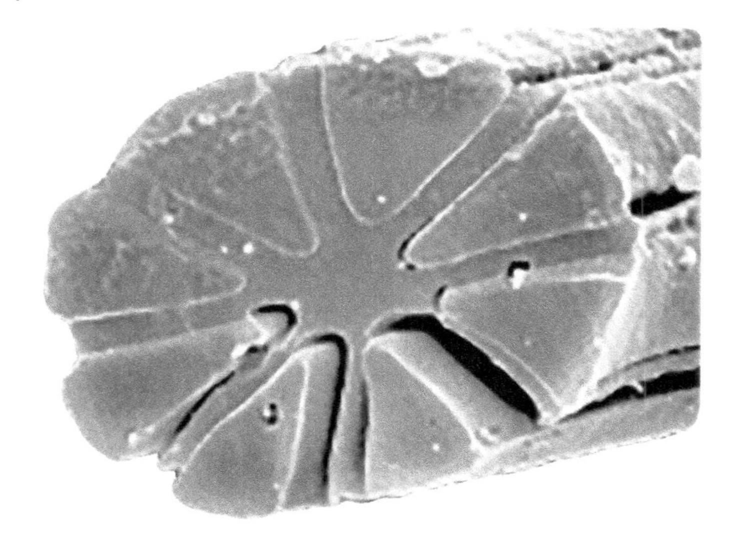

Figure 5.32 Cross-sectional view of split-type microfiber consisting of the star-shaped core and wedge-shaped pedals. *Source:* Courtesy of Toray Industries.

Figure 5.33 Magnified (3000×) longitudinal view of split type microfiber washcloth showing flower pedal wedges of the microfiber.

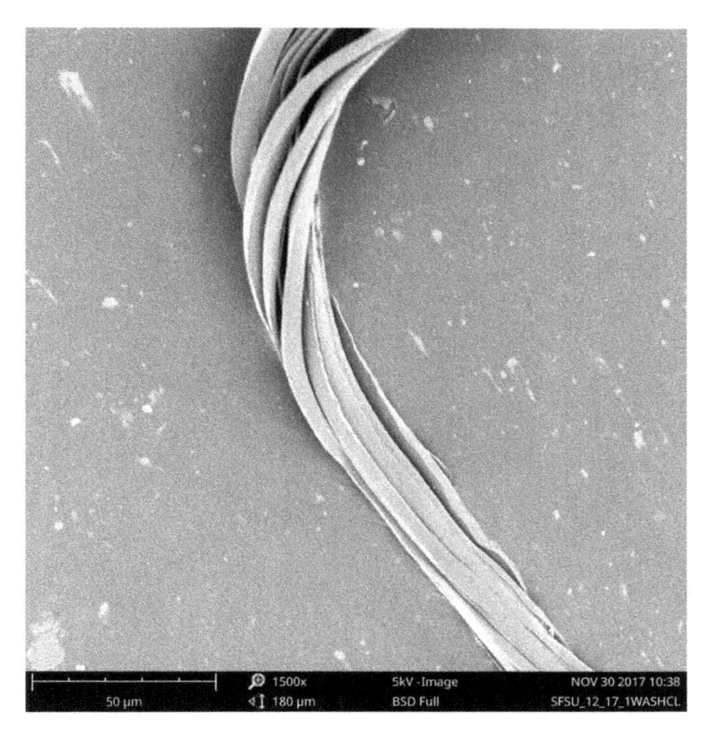

Figure 5.34 Longitudinal view of split-type microfiber washcloth (1500×). Small dust particles fit between flower pedal wedges of the microfiber during cleaning.

bicomponent fiber is specifically engineered for its end use, cleaning cloth; however, it is also used in fashionable products. Figures 5.33 and 5.34 illustrates the split type microfiber in washcloth fabric with 80% polyester and 20% polyamide.

5.17.2 Imitation Leather/Suede

Microfibers in the split form were first developed to create imitation leather. The multiradial bicomponent and multi-islands versions (called Belima-X®) were developed because a microfiber processed using the regular spinning was not sufficient to make imitation leather. In Japan, fiber producers developed their own bicomponent fiber for this use, the multilayer microfiber [7]. Today, imitation suede microfiber fabrics are made by the nonwoven method to create a fiber web material. This material may be impregnated with polyurethane resin to simulate real suede. Figure 5.35 shows where resin covering the microfibers can be easily identified under a 4000× magnification of scanning

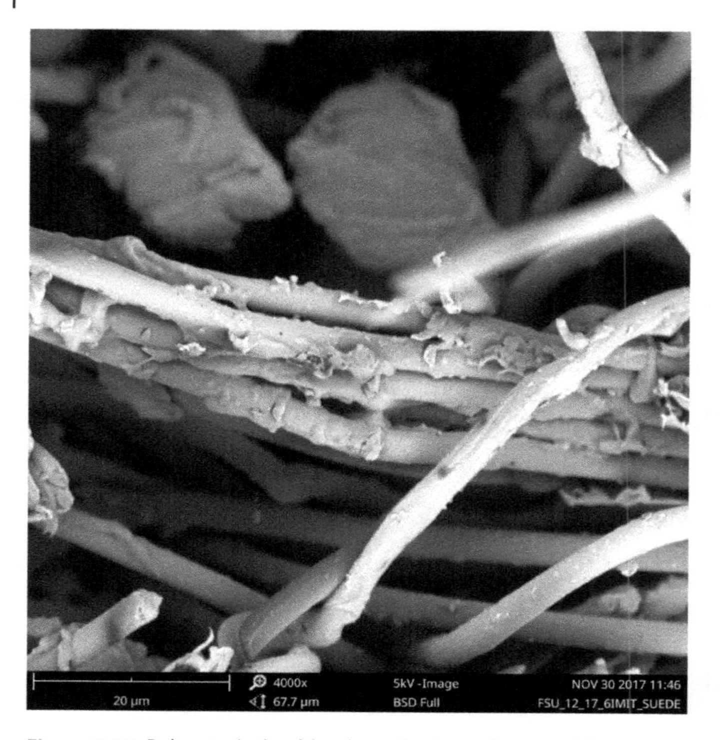

Figure 5.35 Polyester/nylon blend, synthetic suede material – resin-covering microfibers can be identified under 4000× magnification of SEM microscope.

electron microscope (SEM). This suede fabric is a polyester and nylon blend, with the polyester microfibers appearing in the forefront of the picture and regular denier nylon fibers appearing in the background. Fiber web is comprised of entangled fibers, in this case, done by needle punching. Figure 5.36 illustrates a microscopic depiction of fiber entangling of fiber web fabric.

In conclusion, microfibers have had a great impact on the textile industry; they have a presence ranging from underwear and pajamas all the way to outerwear, such as active wear or rainwear. However, recently, environmental reports have stated that microfibers are harmful to the environment. The smaller the fiber, the more the pilling occurs, which can apparently result in fiber breakage. Microfibers break when laundered, and millions of them end up in sea waters and oceans causing pollution [11]. Evidence of polyester, nylon, and acrylic microplastic contaminating shorelines worldwide has been found, which is believed to have come from washing clothes [12]. Scientists have found microfiber pieces in fish that are consumed by humans.

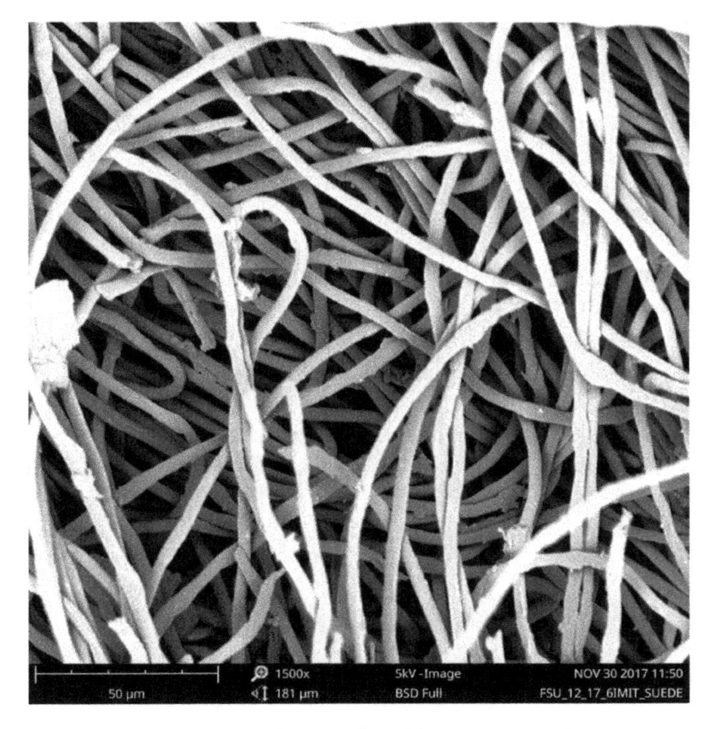

Figure 5.36 Imitation suede microfiber fiber web entangling (1500×).

References

1 Kadolph, S.J. and Langford, A.L. (2002). *Textiles*, 9e. Upper Saddle River, NJ: Pearson Education.
2 Omeroglu, S., Karaca, E., and Becerir, B. (2010). Comparison of bending, drapability and crease recovery behaviors of woven fabrics produced from polyester fibers having different cross-sectional shapes. *Textile Research Journal* 80 (12): 1180–1190.
3 Bueno, M., Aneja, A.P., and Renner, M. (2004). Influence of the shape of fiber cross-section on fabric surface characteristics. *Journal of Materials Science* 39 (2): 557–564.
4 Babaarslan, O. and Haciogullari, S.O. (2013). Effect of fiber cross-sectional shape on the properties of POY continuous filament yarns. *Fibers and Polymers* 14 (1): 146–151.
5 Karaca, E., Omeroglu, S., and Becerir, B. (2015). Effects of fiber cross-sectional shapes on tensile and tearing properties of polyester woven fabrics. *Tekstil ve Konfeksiyon* 25 (4): 313–318.

6 Masson, J. (1995). Apparel end uses, Chapter 10. In: *Acrylic Fiber Technology and Applications* (ed. A. Lulay), 313–340. New York, NY: Marcel Dekker, Inc.
7 Murata, T. (1993). Polyester super-microfibers and their use in apparel textiles. *International Textile Bulletin: Yarn and Fabric Forming* 42–47.
8 Johnson, I., Cohen, A.C., and Sarkar, A.K. (2015). *J. J. Pizzuto's Fabric Science*, 11e. Fairchild Books, Bloomsbury Publishing.
9 Humphries, M. (2008). *Fabric Reference*, 4e. Upper Saddle River, NJ: Pearson Education, Inc.
10 Matsui, M. (2009). The spinning of highly aesthetic fibers. In: *Advanced Fiber Spinning Technology* (ed. T. Takajima, K. Kajiwara and J.E. McIntyre), 115–129. Woodhead Publishing Limited.
11 Le, K. (2017). Microfiber shedding: hidden environmental impact. *AATCC Review* 17 (5): 30–37.
12 Browne, M.A. et al. (2011). Accumulation of microplastic on shorelines worldwide: sources and sinks. *Environmental Science and Technology* 45: 9175–9179.

6
Nanofibers

6.1 Nanotechnology in Textiles

The advancement of textiles is marked by inventions; the first synthetic fiber in the twentieth century, the later development of microfibers, and now, in the twenty-first century, the development and application of nanofibers. The field of nanotechnology has had an impact on many industries, such as material science, optics, medicine, plastics, aerospace, and as well as the textile and apparel industry. Nanofibers exist on a smaller scale than regular fibers or microfibers, with diameters ranging from 1 to 100 nm [1] (see Figures 6.1 and 6.2). Nanofibers are extremely thin and fine fibers. Molecules are said to be the smallest particle, and in comparison to a nanofiber, a molecular chain has a diameter smaller than 1 nm, which is close to the size of a nanofiber [1]. The interest in nanofibers is due to their unique properties, which are basically improvements in chemical, physical, and biological properties due to their nanoscale size [2].

Nanofibers have been utilized in the textile industry in the form of fabric coatings. One of the advantages of a nanocoating for water repellency, when compared to a regular water repellency finish, is that the nanocoating does not compromise the inherent fabric properties, such as the feel and softness, as it does with a regular finish. The consumer does not even know there is a nanocoating applied to their fabrics because such a small scale does not affect the fabric itself. With the help of nanotechnology, textile fabrics gain a new function and thus become multifunctional. Some of these special functions include water and stain repellency, and textiles with more protective functions including UV-protection, anti-odor, and antibacterial properties are also manufactured. Polymer science researchers are interested in the development of high-performance fabrics whether for textile applications for everyday consumers or for military, aerospace, and or medical textiles.

Textile Fiber Microscopy: A Practical Approach, First Edition. Ivana Markova.
© 2019 John Wiley & Sons Ltd. Published 2019 by John Wiley & Sons Ltd.

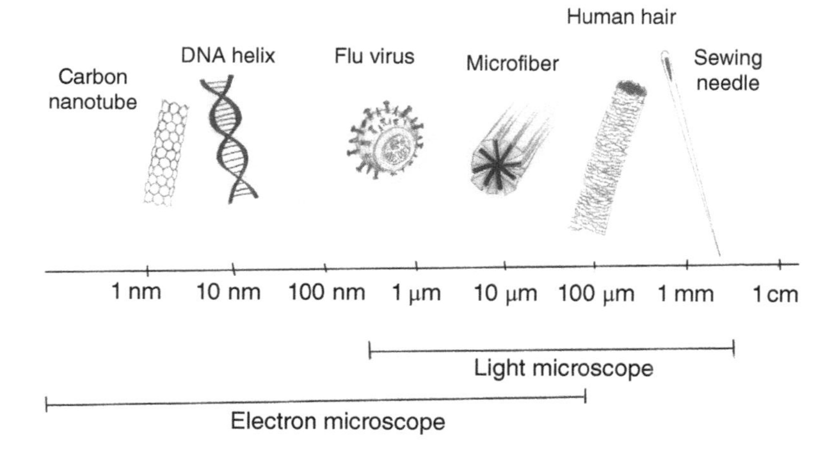

Figure 6.1 Fiber size comparison – regular denier fibers, microfibers, and nanofibers.

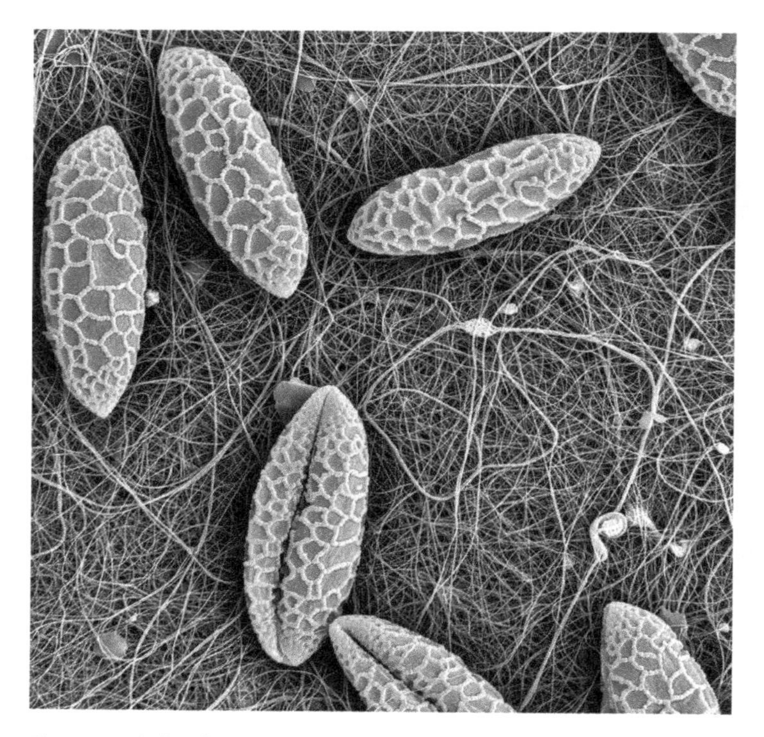

Figure 6.2 Pollen from lily flower on nanofibers (750×). *Source:* Courtesy of Eva Kuzelova Kostakova. Technical University of Liberec, Liberec, Czech Republic.

6.1.1 Production of Nanofibers

Although there are a variety of production methods for nanofibers, two methods are used most frequently: electrospinning and the islands-in-the-sea method. The method of fiber production through electrospinning entails making fibers from a polymer solution through electric force. The polymer is pumped through the very fine tip of a syringe toward a conductive collector plate. High electrical voltage of around 15–25 kV is applied between the tip of the syringe and the collector plate [1]. When a high enough voltage is applied to a droplet, the droplet becomes charged and stretches creating a fiber. If no electrical voltage is applied, the polymer solution would simply fall to the collector plate [1]. One issue regarding electrospinning is its low production speed. A new and groundbreaking method, Nanospider™, was developed using needleless electrospinning technology to make nanofibers.

6.1.2 Uses of Nanofibers

Nanofibers have been used in materials made for the medical care industry such as wound dressings, tissue scaffolds, and medical implants [1] (see Figure 6.3). Nanofiber technology is in its prime developmental stage, and so far scientists are still discovering ways to produce nanofibers in large quantities. This is the reason why their usage is focused on only small-scale items in medical field rather than large-scale items. Small-scale fibers can be engineered in ways that large fibers cannot, such as in medical tissue scaffolds where they have the ability to simulate the extracellular matrix in natural tissue [1]. This is one example of the greater urgency to use nanotechnology in the medical field than in any other field.

One of the uses of nanofibers to create a multifunctional textile is to create a drug delivery fabric – a fabric that will self-release drugs into a person's body merely by wearing it. In recent years, there has been great interest in this biomedical application. Researchers have developed a variety of nanofiber techniques, which can perform this undertaking. It is a fascinating idea that our clothing might also function as a drug delivery system. Medical professionals see many advantages to this system as drug delivery could be localized. One way of achieving this is to create an electrospun nanofiber that would encapsulate the drug agent inside the fiber's hollow core. The core shell structure of these drug-delivery fibers is highly porous through which the drug can be released. A different way of achieving this is to mix the drug substance with the polymer and electrospun together to form the encapsulated fibers. Drug release during both methods is quite constant; however, during the first method, it begins with an initial burst [3].

The reason for interest in these fabrics is their multiple benefits. Currently, drug delivery methods, such as pills and injections, are not as controllable and involve the risk of overdosing. Because a drug delivery fabric constitutes a slow drug release system from a fabric through the skin of the wearer, the risk of an

Figure 6.3 Porous fiber surface is used in nanofiber technologies for the development of knee bone tissue material. *Source:* Courtesy of Eva Kuzelova Kostakova. Technical University of Liberec, Liberec, Czech Republic.

overdose is reduced. One might say that drug delivery textile garments could be a solution to the danger of drug overdoses. It would also be beneficial for small children who despise taking medications, for older individuals who are suffering from dementia and are forgetful about taking medication, or for people who have problems swallowing [4].

6.1.3 Nanowebs

Nanofiber webs have also been used for slow-release drug delivery with its micropore structure that holds drugs temporarily [1]. Nanofiber webs, which are considered a new kind of nonwoven material, are created using the electrospinning method. Nanofiber webs have been successfully used for filtration (see Figure 6.4). Nanofibers can be made of various sizes, as small as 40 nm, and various nanofiber web densities are possible [5]. A thin web is applied onto

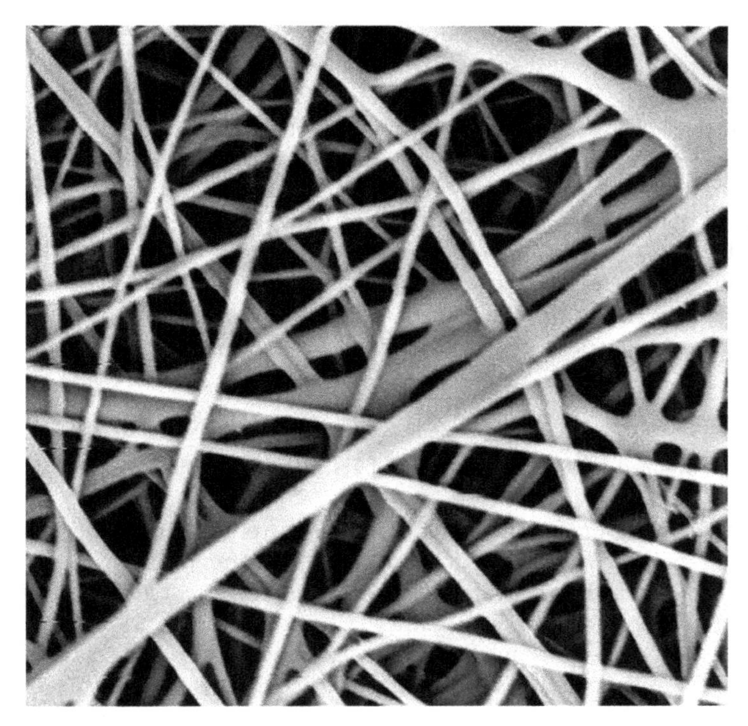

Figure 6.4 Nanoweb electrospun nanofibers from polyvinyl alcohol (water-soluble synthetic polymer)/water solution by needle-less electrospinning technology. *Source:* Courtesy of Eva Kuzelova Kostakova. Technical University of Liberec, Liberec, Czech Republic.

other fibers called substrates, which may be made of polyester, nylon, or other fibers. The nanoweb itself does not have good mechanical properties and therefore is applied on other fibers, serving as substrates, with good mechanical properties [5]. Thus, the substrates provide complementary functionality to the nanofiber web [5].

Nanowebs are used for filters in different applications, such as for water or air, as their highly porous structure and high surface area give them the ability to filter submicron particles. Nanoweb filters can be also used as protective clothing in military uniforms to protect against biological or chemical elements.

Teijin, a Japanese manufacturer, is a leader in nanotechnology. Under their Nanofront® trademark, they have developed an ultrafine polyester nanofiber with a diameter of merely 700 nm. The Japanese manufacturer received a Life Nanotechnology Award in 2016. Nanofront has been using the island-in-sea composite spinning technology to make facemasks and skin care items for the health-care industry. They have also created the technology for creating a continuous yarn of carbon nanotubes (by Teijin Aramid).

An island-in-sea textile helps to trap dust or other dirt particles in between the islands and the sea. Nanofibers of 700-nm diameter are smaller than fine dust particles and oil. These textiles are very effective in wipes for cleaning as they can clean micron-level thick, oily residue. Because of the high surface density of Nanofront fabric, the fabric surface produces a large frictional force that gives the fabric a strong gripping power. This property is highly valued in high-performance products such as hiking gloves which stay on when sweating or in rainy conditions, socks which do not fall off or slip inside shoes, or bedsheets which do not come off the mattress.

6.1.4 Nanocoatings

Nanotextiles are created by taking ordinary fibers and implanting them with nanoparticles.

Common nanocoating or nanofinishing techniques are used in textiles for water and stain repellency. The coating entails the application of small particles resembling whiskers, thus called nanowhiskers, which do not allow water droplets to pass through the fabric but rather remain on the surface making these whiskers waterproof. The same principle used for water repellency is also used in stain repellent nanocoatings. Nanowhiskers also increase the density of the fabric.

Another popular nanocoating technique makes textile fabrics antibacterial or antimicrobial to eliminate odors on clothing. The coating to make textiles odor-free uses silver nanoparticles that kill off any bacteria on contact or prevent bacteria from growing (see Figure 6.5). Silver's antibacterial property has been recognized and used throughout history. Today, it has become even more important as many bacteria are increasingly becoming resistant to antibiotics [6]. The use of silver ions (Ag^{+}) is very powerful, and its antimicrobial activity is efficient against 650 trends of bacteria [6]. Silver is used for many other products such as cosmetics, because of its antimicrobial property. Silver nanocoating is widely used in clothing such as sportswear, camping outfits, socks, and undergarments.

New methods of applying the nanocoatings onto fabrics have been in development. For example, in one method, silver particles in a polymer thin film is then directly applied to a fabric via a layer-by-layer deposition method, also called polyelectrolyte multilayers (PEMs), entailing successive dipping of a substrate in dilute solutions of oppositely charged polyelectrolytes [7].

Nanotechnology is an especially exciting field in textiles because new developments are emerging every year. When a method of applying nanocoatings to fabrics, such as PEM, is developed, further research and analysis is needed to determine on which fabrics it can be used. For example, when this PEM

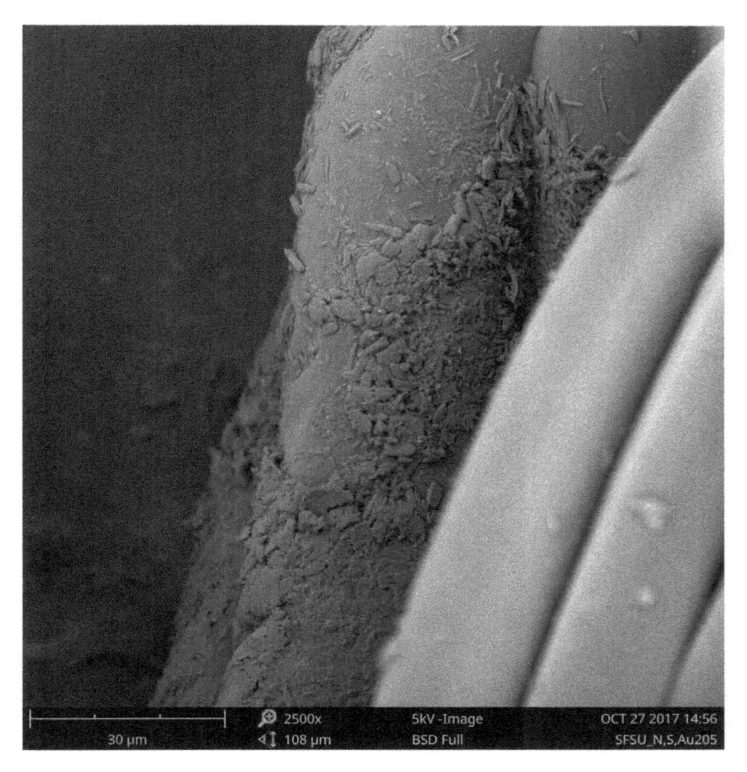

Figure 6.5 Synthetic fibers (nylon/spandex) covered with silver (Ag^+) nanoparticles, giving the fabric antimicrobial finish, odor-free textiles (2500×).

nanocoating method was applied on a natural fabric, silk, and synthetic fabrics, nylon, it worked better on the natural silk fiber than on a synthetic nylon fiber. The silver particle coating resulted in an 80% reduction of bacteria on silk fibers and a 50% reduction of bacteria on nylon fibers [6].

6.1.5 Nanoparticles

Incorporating nanoparticles into textile products will transform ordinary products into products with special functionalities. As previously mentioned, incorporating silver nanoparticles will result in a textile fabric that is antibacterial and antifungal. Titanium dioxide nanoparticles are used to create a textile offering UV protection. Zinc oxide nanoparticles are good candidates for many added functionalities such as zinc oxide nanowires to help produce electricity, which is generated by rubbing the textile's nanowires together [8]. Another nanomaterial worth mentioning is the silicon dioxide nanoparticle. When this nanomaterial is applied to a cotton textile, it enhances the cotton's mechanical

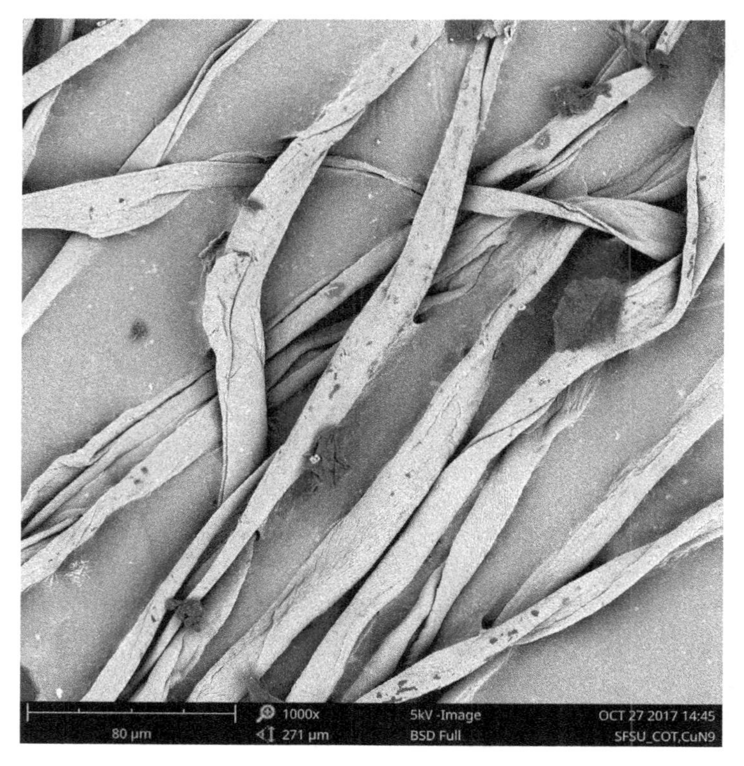

Figure 6.6 Organic cotton coated with nanoparticles – flecked with copper throughout the fibers, giving the end product antimicrobial property (1000×).

properties such as tensile and tearing strength [9]. Copper can also be used for antimicrobial, antibacterial, and antiinflammatory properties. See Figure 6.6 for subtle flecking of *copper* throughout the cotton fibers.

6.1.6 Electrically Conductive Fibers

Scientists have been trying to develop ways to make electrically conductive fibers, and several different methods have been developed, such as using very fine metal wires and inherently conductive fibers, coating textile fibers with an electrically conductive layer, and the addition of conductive particles into a fibrous polymer matrix [1].

With the invention of graphene, there are more opportunities to construct a garment that conducts electricity. Graphene is a tiny, atomic-scale hexagonal lattice made of carbon atoms, which has unusual properties. For its tiny size, it is incredibly strong, much stronger than steel, it very easily conducts electricity, and it can be made into a transparent film. In 2010, two scientists, Andre Geim and Konstantin Novoselov, won a Nobel Prize in Physics for discovering graphene. Because of its tiny size and electrical conductivity, many scientists recognized an opportunity to apply it to clothing via nanotechnology [10]. Conductive textiles are desirable for a variety of purposes and have already been used in medical textiles or work-out outfits to measure muscle movement or heart rate. The creation of conductive textiles with the use of graphene is in the developmental phase and has a promising future. Scientists have even been able to attach graphene-based inks to cotton fibers [10].

Electronic inks made of silver have also been applied to fabrics to make electronic circuits. These inks are also stretchable, which makes the garment more comfortable to wear. DuPont has developed smart textiles with the use of these inks, which will deliver important biometric data such as heart and breathing rates, muscle tension, and form awareness. These smart fabrics are used for high performance wear. See Figure 6.7 for carbon nanofibers with unique fiber shape.

6.1.7 Porous Surface Fibers

A porous fiber surface is used in nanofiber technologies. Porous nanofibers were developed at the Technical University of Liberec, in the Czech Republic, which were to be used for the manufacture of scaffolds for tissue engineering in the medical field. They are made via a needle electrospinning technology. These porous nanofiber fibers are currently used for the development of knee

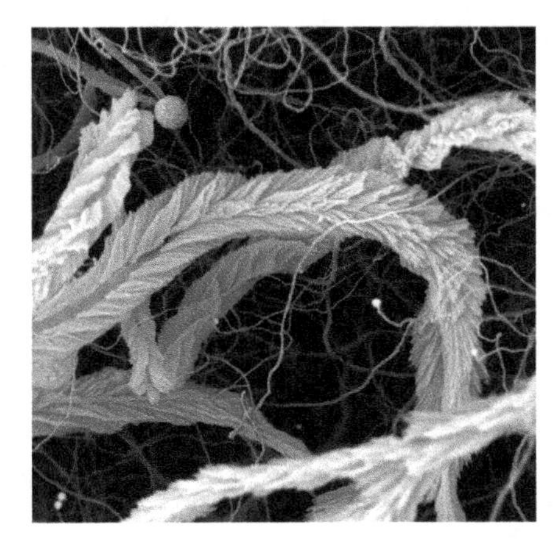

Figure 6.7 Carbon nanofibers possess potential electrical conductive properties. *Source:* Courtesy of Eva Kuzelova Kostakova. Technical University of Liberec, Liberec, Czech Republic.

bone tissue material. The pores act as an absorbent where human cells easily absorb into their surface. These nanofibers are biodegradable (see Figure 6.3).

A porous surface structure makes fibers absorbent. Synthetic fibers lack this property; this is one way it could be created. Moisture is absorbed through pores leaving the fiber surface dry. The pores are basically many voids, small opening in a fiber. In the 1990s, Teijin, the Japanese fiber manufacturer, developed a porous regular denier polyester fiber called WELLKEY with a hollow core. This core is believed to be hydrophilic, water absorbing, and very effective in the removal of perspiration from clothing, such as sportswear.

6.1.8 Microscopy

To view textile fibers on a nanoscale level, an ordinary light microscope will not be sufficient. One might think why not just use higher magnification on a light microscope, but one is not able to see things that are smaller than light's wavelength, which ranges from about 400 to 700 nm. Anything smaller than this cannot be viewed with a light microscope. Nanoparticles are smaller than the light's wavelength, about 10 nm, more or less, while nanomaterials vary in size from 1 to 100 nm.

As optical microscopy only observes on a micron level, to observe nanofibers on a nanolevel, other, more sophisticated methods of microscopy must be used, such as transmission electron microscopy (TEM), scanning electron microscopy (SEM), or scanning tunneling microscopy (STM) [2]. The electron microscope was developed by Ernst Ruska in 1938 and won for him a Nobel Prize in Physics 50 years later. The electron microscope greatly enhanced the ability to explore materials on a different level. The same year that Ruska won his Nobel Prize, Gerd Binning and Heinrich Rohrer also won a Nobel Prize in Physics for their invention of scanning tunneling microscope, which provides three-dimensional images at an atomic level. Though both of these microscopes work on different principles, they both provide highly magnified images. These more sophisticated microscopes are used differently than an optical microscope. An SEM does not have a stage for slides but has microscope chambers where samples are placed. Second, they do not have oculars but a built-in computer and viewing screen. The microscope resembles a kind of computer software through which one can zoom in with a help of a computer mouse. Zooming in and obtaining a precise measurement of the fiber size can be done by merely a click of a button.

References

1 Mather, R.R. and Wardman, R.H. (2011). *The Chemistry of Textile Fibers*. Cambridge: Royal Society of Chemistry.

2 Joshi, M., Bhattacharyya, A., and Ali, W. (2008). Characterization techniques for nanotechnology applications in textiles. *Indian Journal of Fibre and Textile Research* 33: 304–317.
3 Yang, Y.Y., Shi, M., Goh, S.H. et al. (2003). POE/PLGA composite microspheres: formation and in vitro behavior of double walled microspheres. *Journal of Controlled Release* 88: 201–213.
4 Breteler, M.R., Nierstrasz, V.A., and Warmoeskerken, M.M.C.G. (2002). Textile slow-release systems with medical applications. *AUTEX Research Journal* 2 (4): 175–189.
5 Grafe, T. and Graham, K. (2002). Polymeric nanofibers and nanofiber webs: a new class of nonwovens. International Nonwovens Technical Conference, Atlanta, Georgia.
6 Dubas, S.T., Kumlangdudsana, P., and Potiyaraj, P. (2006). Layer-by-layer deposition of antibacterial silver nanoparticles on textiles fibers. *Colloids and Surfaces A: Physicochemical and Engineering Aspects* 289: 105–109.
7 Hyde, K., Rusa, M., and Hinestroza, J. (2005). Layer-by-layer deposition of polyelectrolyte nanolayers on natural fibres: cotton. *Nanotechnology* 16: 422–428.
8 Som, C., Wick, P., Krug, H., and Nowack, B. (2011). Environmental and health effects of nanomaterials in nanotextiles and façade coatings. *Environment International* 37 (6): 1131–1142.
9 Patel, B.H., Chaudhari, S.B., and Patel, P.N. (2014). Nano silica loaded cotton fabric: characterization and mechanical testing. *Research Journal of Engineering Sciences* 3 (4): 19–24.
10 Ren, J., Wang, C., Zhang, X. et al. (2017). Environmentally-friendly conductive cotton fabric as flexible strain sensor based on hot press reduced graphene oxide. *Carbon* 111: 622–630.

7

Recycled Fibers

7.1 Fiber Recycling

The clothing industry generates a great amount of waste, which is not consumer recyclable. Clothing in poor condition, such as a torn pair of polyester pants, cannot simply be thrown in a recycling bin along with other discarded goods such as plastics, cans, and paper. Because there is no way for consumers to recycle textiles, discarded clothing will most likely end up in a landfill.

Today, textile manufacturers are attempting to address the issue of recycling textiles, using terms such as recycled, repurposed, or reprocessed. There is a great interest in developing recycling methods for synthetic fibers (to reduce environmental impact) such as polyester and utilize it on a large scale. Currently, there are two basic types of polyester textile fibers, recycled polyester fiber and virgin polyester fiber that has not been recycled.

Virgin polyester fiber has a wide range of applications. The application of recycled polyester fiber, however, depends on the method of recycling used. The two dominant polyester recycling methods are chemical recycling and mechanical recycling. The end goal is to produce a quality of recycled polyester fiber equivalent to that of virgin polyester fiber, resulting in greater consumer satisfaction. However, the quality of recycled polyester fiber also depends on the recycling method: chemical and mechanical recycling methods.

7.2 Recycled Polyester via Chemical Recycling

The synthetic fiber most utilized by consumers is polyester, which has been recycled by a variety of manufacturers. Polyester recycling is accomplished by a method called *chemical recycling*. Although mechanical recycling (discussed next) has been the predominant recycling method for many years, chemical recycling has been gaining popularity. Teijin, a Japanese manufacturer, offers advanced

Textile Fiber Microscopy: A Practical Approach, First Edition. Ivana Markova.
© 2019 John Wiley & Sons Ltd. Published 2019 by John Wiley & Sons Ltd.

solutions for polyester fiber recycling. In 2000, Teijin developed an environmentally friendly, closed-loop system for chemically recycling polyester. It is called ECO CIRCLE™ and was the world's first technology for polyester fiber chemical recycling. This process includes the collection of polyester items, chemically decomposing them, and in turn, converting them into polyester raw material. This recycled polyester is believed to be comparable to virgin polyester directly derived from petroleum. New recycling methods for synthetic fibers are designed to substantially reduce energy consumption and carbon dioxide emissions compared to regular polyester manufacturing processes.

The quality of chemically recycled polyester increases when it is broken down into a smaller component. For example, the quality increases, and it is more similar to virgin polyester when the chemical recycling involves breaking it down into monomers as opposed to oligomers[1]. Chemical recycling back-to-oligomer results in polymers similar in quality to virgin polyester except for its dyeability. The dyeability in back-to-oligomer chemical recycling is of a lower grade [1].

7.2.1 Microscopic Appearance

As the quality of recycled polyester fiber via chemical recycling is equal to virgin polyester, the recycled fiber is indistinguishable under the microscope. See Figures 7.1 and 7.2. Today, any fiber shape that is produced with virgin polyester can also be produced with recycled polyester (Tomomi Okimoto, personal email communication with Teijin Frontier Co., Ltd., Fiber specialist, 10 July 2018). Therefore, fiber experts would not expect to see differences in these two fibers under a microscope (Diane Irvine, personal communication with US customs specialist San Francisco California, 22 March 2017). Blends of virgin and recycled polyester fabrics are common today. Because chemical fiber recycling is more expensive than making virgin polyester, the recycled fiber is usually blended with virgin polyester to make environmentally friendly products (see Figures 7.3 and 7.4). Some consumers are willing to accept the increased cost of these products.

7.3 Recycled PET via Mechanical Recycling

Recycled polyester fibers have been gaining in popularity in recent years among both consumers and producers. Polyethylene terephthalate (PET) recycling involves taking used plastic bottles and recycling them into a textile fiber. Recycled polyester fabrics are used in apparel, home furnishings, and other products. The high cost of PET recycling results from the amount of energy required to break down the polymer into its chemical parts [1]. Because the

[1] Monomers make up an oligomer and oligomers make up a polymer.

Figure 7.1 Longitudinal view of chemically recycled polyester ECO CIRCLE (1500×).

process of PET recycling is costly, producers do not always have the means to provide recycled polyester fibers in a wide range of fiber types to consumers. Fiber types composed of virgin polyester offer a greater variety than those composed of recycled polyester.

Mechanical recycling, commercialized in the 1970s, is not a new method of polyester recycling. It is a method of recycling via melt extrusion. Mechanical recycling is a multistep process in which collected bottles are separated based on color and, often, the region of origin. The bottles are cleaned, and any outside materials, such as labels and plastic caps, are removed. The bottles are then ground up into small flakes, washed, and dried again. The dried flakes are then melted and extruded [2]. The variety and heterogeneity of the PET waste is the main challenge for mechanical recycling. The contamination and complexity of the PET waste make mechanical recycling difficult [3]. If the feedstock is homogeneous, then it will create high-quality polyester fibers. Mechanically recycled polyester is readily available for manufacturers. In apparel, it has mainly been used in fleece products [4] and nonwovens. Other applications of semimechanical

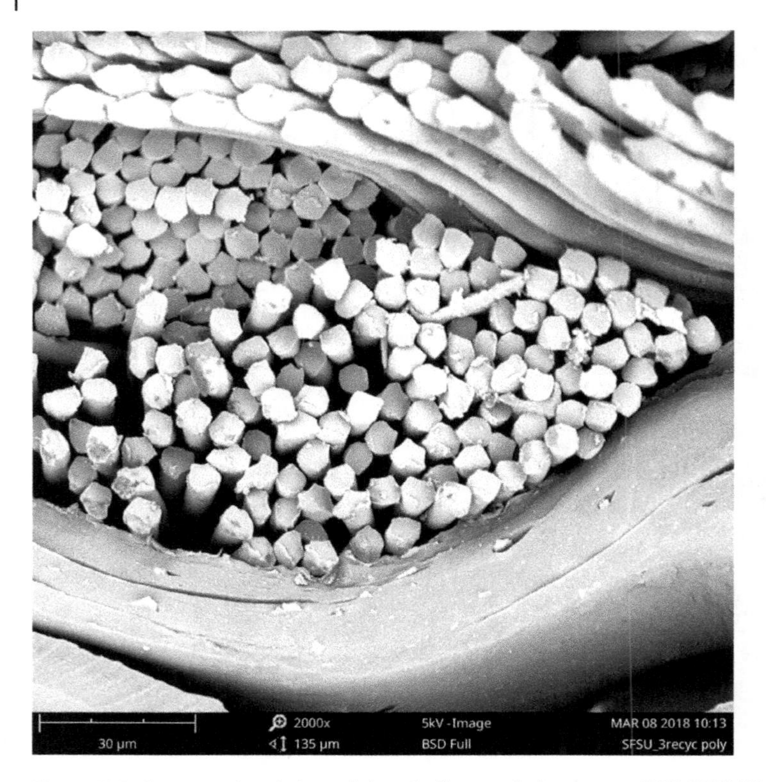

Figure 7.2 Cross-sectional view of chemically recycled polyester ECO CIRCLE irregular round cross-section (2000×).

recycling are found in footwear, bags, and technical textiles. It is interesting to note that it takes 5 plastic bottles to make a T-shirt and 20 plastic bottles for the filling of a winter jacket [2].

7.3.1 Microscopic Images

The quality of recycled polyester via the mechanical recycling is not always comparable to that of virgin polyester (see Figures 7.5 and 7.6). Therefore, mechanically recycled polyester and virgin polyester in a blended fabric could be somewhat distinguishable under a microscope.

First of all, color may be an indicator of a different fiber type because mechanically recycled polyester is difficult to dye, i.e. has poor dyeability. Therefore, mechanically recycled fibers will be of dark, dull colors similar to the colors of the plastic bottles of which they were made. On the other hand, the virgin polyester fibers in the blend could be of any desirable color, and are usually used to enhance the appearance of the fabric.

Figure 7.3 Longitudinal view of 50% chemically recycled and 50% virgin polyester blend ECO CIRCLE (2000×).

7.4 Recycling Nylon

There have been developments in nylon recycling especially in auto industry and in composite materials. Nylon is a fiber that can also be recycled, especially nylons used in carpeting. However, there are companies around the world that recycle nylon fibers not only from carpeting but also from fishing nets (ECONYL®) and preconsumer waste (REPREVE®). Some of these recycled nylons are used for carpeting (ECONYL) and some for textile products (REPREVE). ECONYL, recycled nylon 6 has a unique clover-like cross-sectional shape to help preserve carpet buoyancy. However, recycled synthetic nylon fibers can have a variety of cross-sections including trilobal just as their virgin nylon counterparts.

7.5 Recycled Cotton

Recycling cotton, the second most commonly used consumer fiber for clothing, will be discussed next. New methods for cotton recycling are also designed to substantially reduce the use of water and pesticides, which the

Figure 7.4 Cross-sectional view of chemically 50% recycled and 50% virgin polyester blend ECO CIRCLE depicting irregular fiber cross-section (2000×).

growth of cotton requires. Recycling cotton fibers into a new material saves a substantial amount of water, around 70%. Cotton production, known for its large water footprint, is predicted to become unsustainable in the future. With the advent of global droughts, water is becoming scarce, especially in China and India where significant acreage is dedicated to cotton production. However, textile recycling also has its drawbacks. For example, cotton fibers are damaged in the mechanical recycling process and become shorter, further shortening with each usage. The resulting fibers are of lower quality compared to its virgin counterparts. As it is difficult to spin short fibers into new thread, they are usually blended with new cotton fibers for strength and quality.

The mechanical recycling process is used for cotton as well as wool. Old cotton fabric, from discarded clothing or manufacturing waste, is collected, sorted, and shredded, producing recycled fibers. During the shredding process, cotton fibers are damaged and become shortened in length. The shortness of the fibers determines the thickness of the yarn. Thick threads can

Figure 7.5 Longitudinal view of mechanically recycled polyester made from beverage bottles – Polartec® fleece 100% polyester (88% recycled, 12% virgin) showing fiber impurities (1500×).

produce only thick fabrics. As a result, there are a greater number of denim products made out of recycled cotton material.

7.5.1 Microscopic Appearance

Because of the damage to cotton fibers occurring during the recycling process, recycled cotton fabrics are less smooth, more hairy, and also weaker in strength compared to virgin counterparts [5]. Recycled cotton fibers also differ in their microscopic appearance from virgin counterparts. Small and large fiber cracks are clearly visible (see Figure 7.7). Viewing the fibers in their full length is recommended (see Figures 7.8 and 7.9).

7.6 Recycled Wool

Mechanical recycling is not only used for cotton but is also used for wool recycling, sometimes referred to as wool processing. Similar to cotton, recycled

Figure 7.6 Cross-sectional view of mechanically recycled polyester – Polartec fleece 100% polyester (88% recycled, 12% virgin) depicting round fiber shape (2000×).

wool is also damaged when shredded during processing and becomes weaker and of lower quality. Because of this damage to the wool fibers during recycling, recycled wool fabrics are not as strong and smooth as their virgin counterparts [5]. Recycled wool products are usually reinforced with recycled polyester or nylon fibers to strengthen the wool.

7.6.1 Microscopic Appearance

As mentioned earlier, because recycled wool fibers are not as strong and as smooth as virgin wool fibers, their appearance differs. When viewing recycled wool under a microscope, one can see overlapping scales which are few and far between. The damage that occurs during the shredding process is clearly visible on the fibers, characterized by fiber cracks and broken off scales (see Figures 7.10 and 7.11).

Figure 7.7 Longitudinal view of recycled cotton fibers depicting small and large fiber cracks (2500×).

7.7 Other Recycling Methods – Using a Rayon Manufacturing Method to Recycle Fibers – A Dissolution-Based Recycling Method

Because of the increase of textile waste and the scarcity of natural resources, companies in the textile industry are always looking for new ways to reuse old or unwanted textiles. VTT, a Finnish research center, has developed a new method to recycle cellulose fibers, mainly cotton. The fibers are dissolved into a cellulose solution and made into a new fiber with the utilization of wet spinning for the first time. The dissolution-based recycling method is very similar to the process used to manufacture rayon fibers. This new fiber is produced using the same technique and the same equipment as that used to manufacture viscose rayon. However, by avoiding the use of carbon disulfide in the dissolving process, the recycling method is more environmentally friendly compared

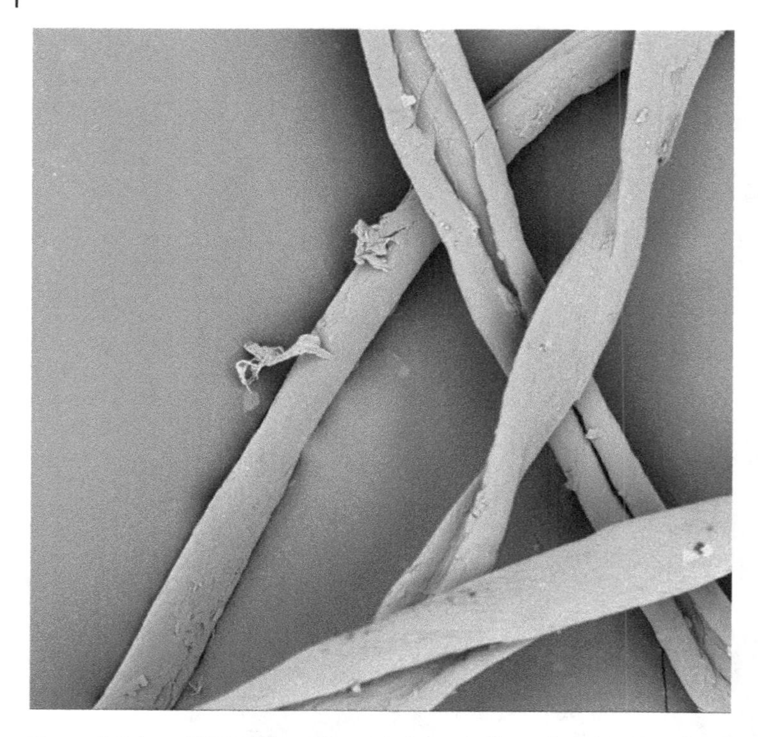

Figure 7.8 Longitudinal view of recycled cotton fibers showing fiber damage (left) (1500×).

to the usual viscose method. This new fiber is to be used for knitted fabrics. Newcell, a Swedish technology company founded in 2012, uses a similar method of dissolving cotton and other fibers, such as rayon, into a dissolved pulp, and then using this to create new 100% cellulosic fibers. The method is similar to how rayon fibers are manufactured using wood pulp or cellulose derived from trees rather than cotton from old clothing.

7.7.1 Microscopic Appearance

Because they are produced using the same technique as that used in manufacturing viscose rayon, under a microscope, these recycled fibers look similar to rayon viscose fibers. Viscose rayon under a microscope is identified by many striations, lengthwise lines. In cross-section, the fiber appears to have serrated edges, resulting from precipitation in an acid bath when extruded from a spinneret.

7.7.2 Recycling Blends

Even though there are few methods of recycling textile fibers with a single fiber content, researchers wish to advance/scale textile recycling so that blended

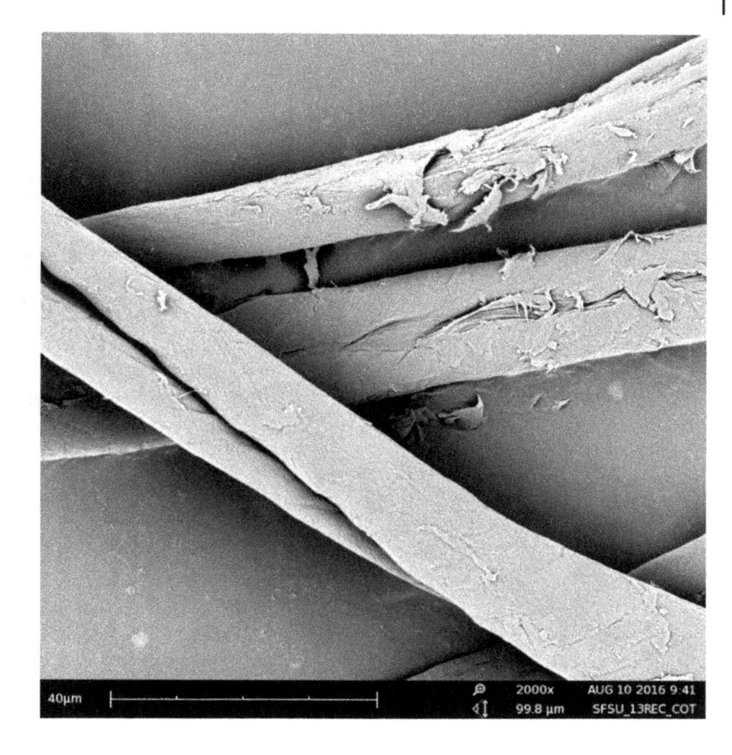

Figure 7.9 Magnified (2000×) view of cotton recycled fibers depicting fiber damage.

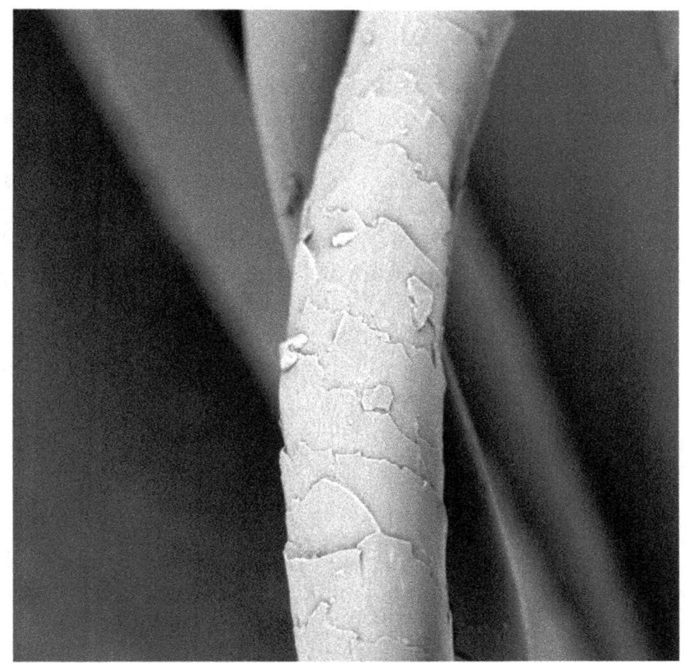

Figure 7.10 Longitudinal view of recycled wool fiber depicting broken off scales (2000×).

Figure 7.11 Longitudinal view of recycled wool fibers showing almost no cuticle scales left on fiber (2000×).

fabrics can also be recycled. For example, it is challenging to recycle a cotton/poly blend because the two fibers are composed of inhomogeneous materials and are tightly bonded. It is not possible to separate the two types of fibers using mechanical separation. However, some advancements now enable the recycling of blended fiber fabrics. A new method has been developed in which the cotton fibers are dissolved via salty (ionic) liquid and later regenerated into recycled fibers. The polyester fibers remained unaffected by the salty liquid and can be separated and recycled by melting them into new textile fibers or plastics [6]. Despite the challenges, new recycling methods to recycle fiber blends are being developed. For example, nylon/spandex fabrics can be recycled by direct melt mixing [7].

References

1 Shen, L., Worrell, E., and Patel, M.K. (2010). Open-loop recycling: ALCA case study of PET bottle-to-fiber recycling. *Resources, Conservation and Recycling* 55: 34–52.

2 KroB, S. (ed.) (2013). New yarns from old bottles: recycling trends and technologies in yarn manufacturing. In: *Fibers and Filaments: The Experts' Magazine*, vol. 16, 13–23.
3 Park, S.H. and Kim, S.H. (2014). Poly(ethylene terephthalate) recycling for high value added textiles. *Fashion and Textiles* 1 (1): 2–17.
4 Gullingsrud, A. (2017). *Fashion Fibers: Designing for Sustainability.* Fairchild Books, Bloomsbury Publication.
5 Natural Resources Defense Council (2012). Fiber Selection: Understanding the impact of different fibers is the first step in designing environmentally responsible apparel. https://www.nrdc.org/sites/default/files/CBD-Fiber-Selection-FS.pdf (accessed 18 September 2017).
6 Byrne, N. and De Silva, R. (2015). Progress in separating, recycling cotton and polyester blends. Industrial Fabrics Association International. https://www.ifai.com/2015/07/15/progress-in-separating-recycling-cotton-and-polyester-blends (accessed 7 October 2017).
7 Lv, F., Yao, D., Wang, Y. et al. (2015). Recycling of waste nylon 6/spandex blended fabrics by melt processing. *Composites Part B: Engineering* 77: 232–237.

8

Historic Fibers

8.1 Textile Fibers and History

Archeological textiles finds are rare. As it is difficult for textiles to survive over long periods of time, most of our information about clothing worn in the past derives from paintings or descriptive texts. Unless special conditions are present, fabrics created from plant and animal fibers (i.e. organic material) degrade in a short period of time. For example, the hot and arid environment of Ancient Egyptian tombs allowed the preservation of many different items for over 5000 years. The tombs, usually sealed against the outside air, provided a desiccated burial environment. In these sealed burial places, an anaerobic environment inhibited the destructive growth of bacteria. In addition to items preserved in sealed tombs, Egypt's hot and arid desert conditions (i.e. a microclimate) also led to the recovery of many archeologically valuable items. Additionally, waterlogged and frozen items have been preserved as the presence of water created an anaerobic environment [1].

When archeologists excavate newly discovered sites, often old graves, if they are lucky, items such as pottery, jewelry, and textiles are present. As textile items require special care, conservation techniques are usually followed in order to preserve the textile and prevent further damage. To properly identify these very rare finds, historical textile fibers undergo a thorough microscopic examination. Excavated items (i.e. material culture) provide historians with a great deal of information about different cultures. Knowledge about the types of fibers prevalent in certain geographic areas of the ancient world is available to scholars. It is important to remember that historians encounter only plant and animal fibers in historic textile excavations as manufactured fibers were not developed until the twentieth century.

We know for certain that in Ancient Egypt, linen fibers were prevalent. In Mesopotamia, wool fibers are believed to have been mainly used. In India, cotton fibers were common, while in China, silk was the preferred fiber. Newly discovered textiles are studied extensively by scientists. One tool used in

Textile Fiber Microscopy: A Practical Approach, First Edition. Ivana Markova.
© 2019 John Wiley & Sons Ltd. Published 2019 by John Wiley & Sons Ltd.

attempting to identify the fibers is microscopic examination conducted in light of our knowledge of contemporary textile fibers. Historians obtain a collection of local animal fibers from sheep, goats, etc., and microscopically compare their excavated fibers with these contemporary fibers. However, in many cases, due to age and the original conditions of their deposition, fibers from archeological textiles are damaged. In addition, archeological textiles are usually contaminated with surface residues. Because of these conditions, archeological textiles are more difficult to identify and more difficult to compare with their modern counterparts. Also, some types of plant and animal fibers, which were utilized in the past, are not used or cultivated anywhere today.

8.1.1 General Information – Ancient Textiles

Students of apparel and textiles learn about the great designs of the Ancient Greeks mainly as evidenced in Greek statues. The most well-known garments worn by women of that era are the Ionic and Doric chitons. These beautifully draped garments are always depicted as flowing, smooth, and somewhat transparent. This fabrication is definitely breathtaking. The two main fiber types used by the Ancient Greeks were linen and wool [2]. While wool was cultivated domestically, the use of linen is believed to have been imported from Egypt or the Middle East. However, many scholars believe that silk and cotton were also used during this period. Due to a lack of evidence, this belief has been disputed.

Next, we will consider genuine evidence of fibers from this period; fibers that are identified by the scientific methods of microscopy.

Newly excavated textile specimens are usually heavily damaged and soiled. Before the specimens are examined under the microscope, they require proper cleaning as attached soil could mask the fiber's characteristics preventing an accurate examination and identification. Excavated animal hair fibers are cleaned in an ultrasonic bath of distilled water with the 2% addition of a neutral pH detergent solution [3]. Specimens must be handled with care so as not to cause further damage during cleaning. Tweezers should be used when removing fibers from a bath for rinsing. Many fibers are weaker when wet and, if yanked without care, can be ripped. For example, wool fibers are weaker when wet, usually losing almost half their strength. Overall, protein fibers are weaker than cellulosic fibers when wet. As the hydrogen bonds in wool are weakened when moisture is present [4], excavated wool specimens must be dried before handling and examination.

8.1.2 Greek Textiles

Greece has many archeological sites enabling historians to learn a great deal about its rich history. However, a review of case studies of excavated textiles in Greece shows that most of these textile artifacts remain unidentified [5]. It is very important for textile historians that the fiber identification of Ancient Greek textiles be conducted with accuracy. Historians reopened a case of previously identified Greek textile fibers to determine whether they could verify that the prior fiber identification was correct [5]. As a matter of fact, they were able to reverse some of the past fiber identifications of excavated textiles found in the Kerameikos cemetery in Athens, Greece. These textile finds, dating back to the fifth century CE, were found in a grave belonging to Hipparete, the granddaughter of Alcibiades, an Athenian General during the Peloponnesian War (431–404 CE). In the past, scientists had tested the textiles for fiber identification yielding somewhat surprising results as silk fibers were found to be present, a very unusual occurrence during that time period in Classical Greece.

As this case was revisited using a state-of-the-art, nondestructive instrumental analysis, an environmental scanning electron microscope (ESEM) which magnifies up to 50 000 times and FTIR – Fourier Transform Spectroscopy, which is an advanced type of spectroscopy which identifies chemical compounds in a sample [5]. FTIR microspectroscopy is a useful tool that identifies the chemical composition of the fiber. The ESEM is a step up from the scanning electron microscope (SEM), see Chapter 1. The high level of ESEM's magnification provided clear morphological images indicating four different cellulosic fiber textiles. Three of the ancient fibers were described as having "have smooth surfaces, cylindrical shape, and exhibiting nodular thickening across their length, all characteristics of cellulosic bast fibers, such as flax" (p. 526). The fourth cellulosic fiber was very different as it lacked the bast fiber morphology and "is flatter, rather than cylindrical, with occasional convolutions along the longitudinal axis of the fibers, characteristics indicative of cotton fibers" (p. 526).

The initial analysis was conducted by SEM, which is suitable for large textile fragments that cannot be taken apart. The ESEM analysis was very instrumental in determining the morphology of the four different fibers from this burial site. The microscopic image was very detailed and provided clear distinctions among the fibers. This time, with the use of advanced ESEM, the fibers were identified as cotton, not silk, which is equally unusual for the Classical Period in Greece (see Figure 8.1).

Another novel scientific method of identifying some bast historic fibers is the observance of crystal formations around the fibers themselves. Researchers note the fibrillary orientation mainly through the use of polarized light microscopes that can clearly detect the presence of calcium oxalate crystals. The presence of crystals around bast fibers has interested several researchers [6, 7]. These crystals are not specific to the fibers from which textiles are made but are commonly associated with many plants. They are a part of the plants' biology. These crystals were unknown to biologists until Leeuwenhoek invented his microscope and started exploring the microscopic world. As you remember from Chapter 1, he was the cloth tradesman who started recording the small organisms that he saw under his simple optical microscope, resulting in

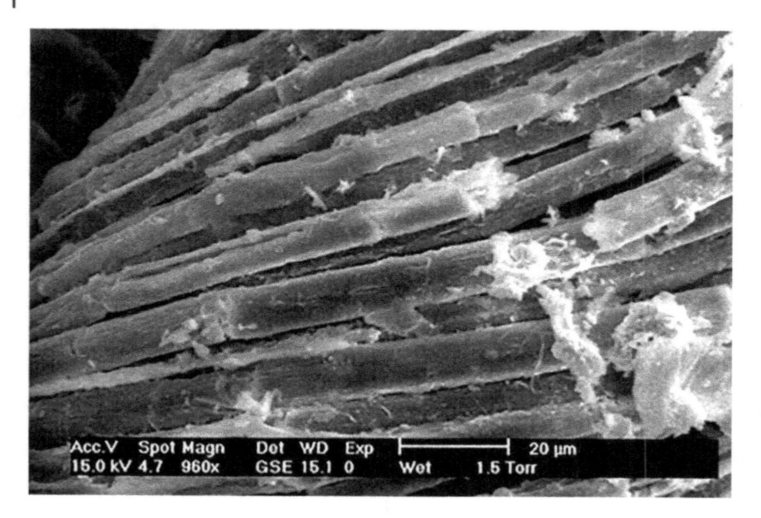

Figure 8.1 Fiber image (Ancient Greek textile) helped archeologists correctly identify misidentified cotton fibers. Fibers appeared to have convolutions which are indicative of cotton fibers not silk as previously believed. *Source:* Taken from figure 6 in Margariti et al. 2011 [5]. Reproduced with permission of Elsevier.

Leeuwenhoek being named "the father of microbiology." He was the first to actually record small calcium oxalate crystals on plants.

The main function of calcium oxalate crystals in plants is to protect the plants from herbivory and to regulate high-capacity calcium (Ca) [8]. The biomineralization process producing the crystals is complex and environmentally derived, resulting in crystals which differ in shape and size depending upon the plant species. These crystals are not in the fibers but are found in the associated fiber tissue. If the tissue is not present because it was removed during the processing (such as retting), then this method of identification cannot be used. The reason why polarized light microscopy (PLM) is recommended is because of the crystals' anisotropic optical properties.

It is a very difficult task to distinguish among bast fibers because many of them have similar microscopic and chemical (except jute) characteristics. Many fiber identification testing methods such as burning and chemical analysis are inconclusive in distinguishing bast fibers. The use of two other methods has been suggested in distinguishing bast fibers (flax, nettle, ramie, jute, and hemp): measuring the fibrillary orientation of fibers through the use of PLM and noting the occurrence of calcium oxalate crystals in bast fibers [7]. Both of these methods require the utilization of microscopy. Bast fiber identification by the presence of crystals has been recommended as an effective method and is discussed next. Two different crystal formations are noted as being present in bast fibers, cluster crystals, and solitary crystals (see Figures 8.2 and 8.3).

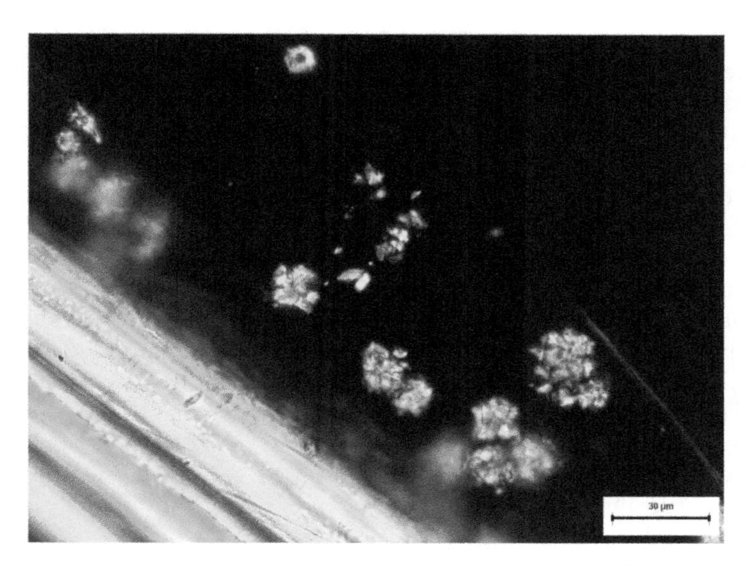

Figure 8.2 Cluster crystals in nettle (also found in ramie and hemp) viewed under polarized microscopy. Note: Flax fibers have no crystals. *Source:* Bergfjord and Holst 2010 [7]. Reproduced with permission of Elsevier.

Figure 8.3 Solitary crystals found predominantly in jute viewed under polarized microscopy. Note: Flax fibers have no crystals. *Source:* Bergfjord and Holst 2010 [7]. Reproduced with permission of Elsevier.

Cluster crystals occur in nettle, ramie, and hemp. Solitary crystals occur in hemp and jute. There are no known crystals in the flax fibers. However, the authors emphasized that the absence of crystals should not be used as an identifying factor (for example for flax) [7]. Crystals could be absent in other fibers for different reasons. Only the presence of crystals should be applicable in the identification of bast fibers.

The procedure to identify the crystals is as follows [7]. A normal mounting procedure is utilized. Clean fibers should be tweezed apart and placed on a slide. A mountant (preferably, for this analysis, galvanol) is applied, and the sample is covered with a cover slip. The fibers are inspected with crossed polars and examined for crystals by rotating the stage to the best possible position to view the contrast. With the crossed polars function, the crystals, if present, should be visible as diamond-like matter, easily distinguished from the fiber itself and in close proximity to the fiber. If solitary crystals are found, then fibers are probably jute, whereas if cluster crystals are found, then the fibers could be nettle, ramie, or hemp. In the case of archeological finds, the crystals may be very distorted or fragmental sometimes making accurate crystal identification impossible. It is recommended that a higher magnification microscope is used, such as SEM (plasma ashing analysis), to further analyze the samples.

Plant fibers have a wall composed of microfibrils that are few nanometers in diameter. These microfibrils are composed of glucose polymer cellulose set in bundles and wrapped around the fiber in spiral-like, helical pattern. Microfibrils are oriented in relation to the length of the fiber. It is important for the textile scientist to know the direction of the microfibrils in fibers because they determine the stiffness and strength of the fibers. However, the direction of the fibrillary orientation also varies among bast fibers and could, therefore, be used as a key in differentiating the fibers [7]. "Most textile fibers show anisotropic optical properties that are determined by the orientation of the microfibrils" (p. 1193). This fibrillar orientation can be determined with the help of PLM. The procedure required to conduct the fibrillar orientation test for identifying bast fibers is discussed next [7]. Once a sample is prepared, the examiner should start with the analyzer parallel to the polarizer. For the best contrast, it is imperative to adjust both the analyzer and the rotating stage. The correct fibrillary orientation will be seen only if the examiner focuses on top of the fiber cell rather than the bottom of the fiber cell, if not the result will be invalid. Two possible results may occur, either a Z-twist in which fibrillary bundles are oriented to the left or an S-twist in which the fibrillary bundles are oriented in the opposite direction to the right (see Figure 8.4). Hemp and jute fibers' fibrillar bundles are oriented to the left, corresponding to the Z-twist; and flax, nettle, and ramie fibrillary bundles are oriented to the right corresponding to the S-twist (see Figure 8.5).

Figure 8.4 Fibrillary bundles orientation. Z-twist – bundles oriented to the left or S-twist bundles oriented to the right.

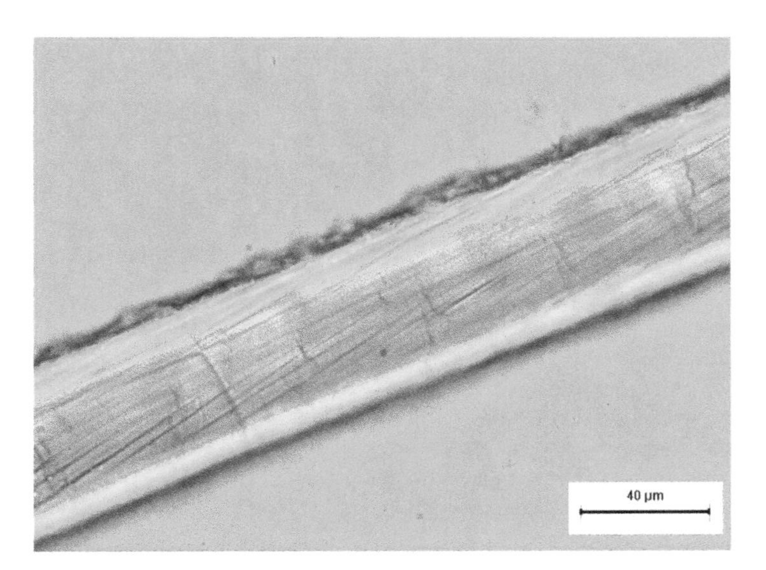

Figure 8.5 Nettle fiber identified by the S-twist direction of the fibrillary structure viewed under polarized microscopy. *Source:* Taken from figure 8 in Bergfjord and Holst 2010 [7]. Reproduced with permission of Elsevier.

8.2 The Use of Hemp in Central Europe

The fiber choice of Eastern European peasants of the Slovak Carpathian Mountains was primarily hemp and wool, both of which were readily available in the geographical region. Eastern European peasants of the nineteenth century were fully self-sufficient. They would not only grow their own food but they would also cultivate and process their own fibers of which they would produce cloths. A ball of hemp thread was found in an abandoned cottage (from c. 1830) located in Slovak Carpathian Mountains. Hemp fiber shape was viewed under an SEM and compared to the morphology of today's hemp fibers. When compared to today's hemp fibers, some differences can be observed.[1] Individual hemp fibers are referred to as fiber ultimates. To form the bundle, hemp fibers are glued together by a binding agent lignin or holocellulose. When viewed historic and today's hemp fibers under an SEM, both are composed of fiber bundles. However, the historic hemp bundle fibers have some fibrous matter (bits of shive) around it,[2] whereas today's hemp fibers do not have the matter around the fiber bundles. See Figures 8.6 and 8.7 for a comparison of new hemp and historic hemp fibers. This difference might be an indication that the historic fiber bundles did not go through as of a rigorous processing as did the new hemp fibers. These bits of shive were probably not removed in the scutching step, which is the mechanical removal of the bits of stalk. Because the historic hemp fibers were found in a yarn ball form, the intended end use of the yarn is not known. It might have been intended to be used for clothing or some other nonclothing items such as bedding or table cloths. Such nonclothing items would not require intricate processing, and they would have had a rougher feel, be thicker, and bulkier.

8.3 Egyptian Textiles

The Ancient Egyptian civilization is one of the greatest in the history of humankind. Because of their strong belief in afterlife, they have left behind in their tombs many objects including textiles. These textile artifacts offer strong evidence of the types of fibers they used for their clothing. The Ancient Egyptians are well-known for their utilization of linen in their everyday garments. Flax plants, from which linen is made, were grown along the Nile River, cultivated for the use by the Egyptians. Linen was a fiber used for the living and as well as for the dead as Egyptian mummies were wrapped in linen cloth. Wool was not a fiber favored by the Egyptians as they perceived it to be as

---

1 Hemp fibers come in fiber bundles.
2 It is shown in purple color on the micrograph.

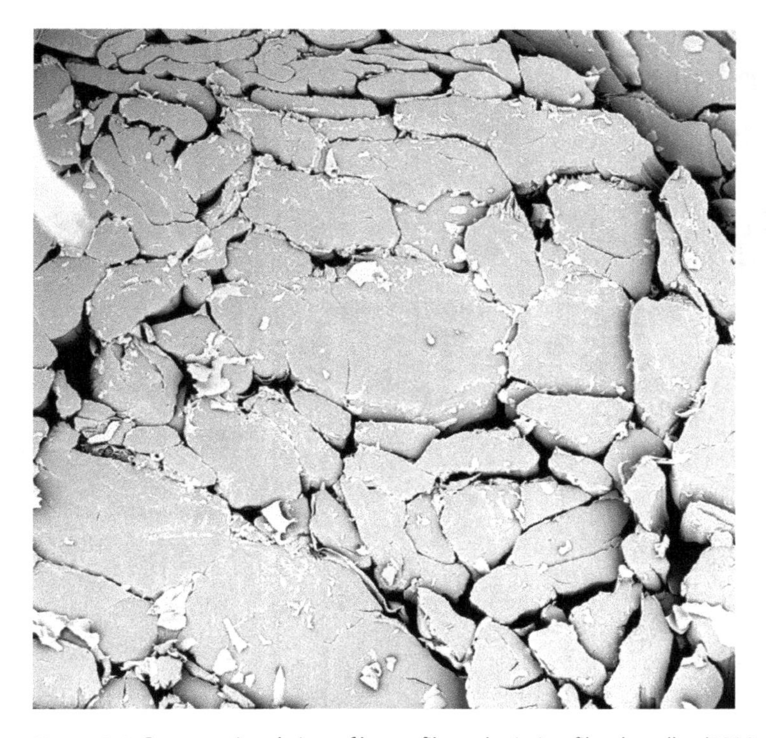

Figure 8.6 Cross-sectional view of hemp fibers depicting fiber bundles (1000×).

unclean, especially when used in temples. However, they did not avoid all animal furs; they liked to wear animal skins such as the leopard. They believed that by wearing these skins they would obtain the animal's powers.

8.3.1 Middle Kingdom Linen Cloth

This linen cloth depicted in Figure 8.8 is from a cloth left in a passage to a burial chamber of a looted tomb of Queen Neferu from the Middle Kingdom Period, Eleventh Dynasty, Thebes, Egypt. This artifact is currently located at San Francisco State University's Global Museum. Because detached fibers were not available for viewing under a light or SEM, only a simple stereo microscope was used to view the cloth's surface. Although fiber morphology cannot be confirmed through this observation, museum experts have indicated it is linen because most of the cloth found in Ancient Egyptian tombs is made of linen fibers [9]. However, there were different spinning traditions for specific fibers and regions knowledge of which can help us confirm Ancient Egyptian origin. For example, Egyptian linens were typically S spun (left direction), and textile fibers from Europe or India were Z spun (right direction) [9]. If you look closely

Figure 8.7 Cross-sectional view of historic hemp fiber bundles, nineteenth century Slovak Carpathian Mountains, depicting bits of shive (in purple) not removed in fiber processing (2500×).

at the yarns, it is clear from Neferu's cloth that the threads were S spun, confirming its Ancient Egyptian origin. Ancient Egyptians had a variety of weave types, but this one appears to be a tabby weave, which is the simplest type of weave where simply the weft thread passes over and under the warp threads [10]. The micrograph depicts the true color (beige) of the linen cloth. Linen fibers do not dye well, and therefore, the Ancient Egyptians would either leave their linen in its natural color or bleach it.

8.3.2 Romano-Egyptian Textiles

Both optical and scanning electron microscopes were used to identify textile fibers excavated from a newly discovered site in Egypt [3]. These fibers were believed to date back to early third and late fifth century BCE. More specifically, these textile fibers are of Roman origin excavated at Karanis, Egypt. Optical microscopes are usually sufficient to identify natural fibers. However, this researcher also used the SEM to recognize distinguishing fiber features at risk

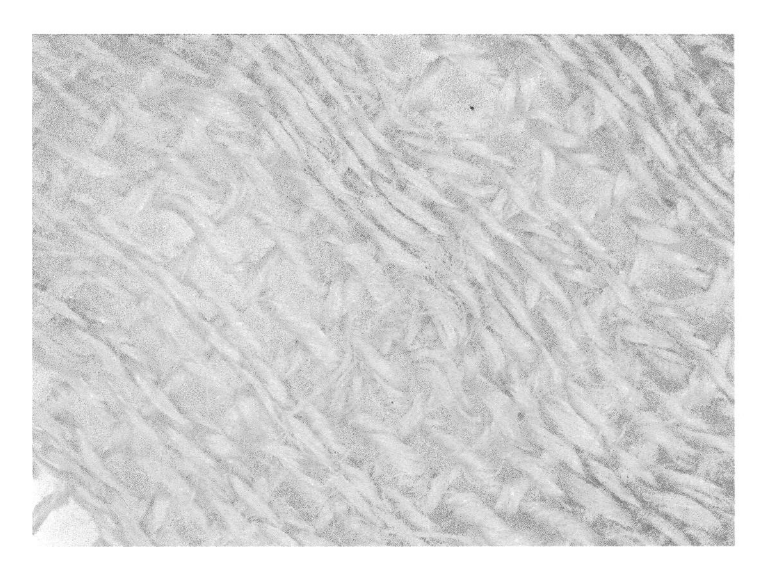

Figure 8.8 Linen cloth from burial chamber of Queen Neferu, Middle Kingdom Period, Eleventh Dynasty, Egypt.

of being damaged and rendered obscure or even obliterated [3]. Therefore, an optical microscopic tool might not be sufficient to analyze these obscure areas more precisely.

Using the optical microscope (400 : 1 magnification) to analyze these obscure areas more precisely, the researcher was able to determine that all three of the different fibers found were of animal hair origin [3]. However, in this case of examining ancient textiles, scanning electron microscope was used to further determine the fiber hair's source. At a 1500 : 1 magnification, the fiber surface details were apparent; and at an even higher magnification of 3500 : 1, fiber features such as the edges of scales could be seen.

Because of their use in modern apparel, much research has been done to identify different hair fibers. Techniques identifying the animal origins from hair fibers through the use of microscope include the anatomic structure of the fibers such as the appearance of the medulla or the shape and size of the cuticle scales. Cuticle scales are analyzed and distinguished in three ways: the scale height, shape, and width. Thus, distinguishing hair fibers from ancient archeological clothing samples through the microscope should be easier. However, if the scales are damaged, measuring their height of the scales becomes more challenging [3]. Another way of determining the animal origin of hair fibers is through examination of the fiber diameter: the smaller the fiber diameter, the better the quality (or the finer the fiber) of the hair fiber. Coarse fibers have the largest diameter of many wool fibers including specialty wools. These are sometimes called kemp fibers. The hair fibers identified from one burial site noted in this study were of wool, goat, and camel hairs. There was a variety of wool hairs that included fine, medium, and coarse diameters. There were many kemp fibers with fiber diameters greater than 50 μm. The researchers were able to distinguish the medullas in these different animal hairs.

It is very interesting to note that with the help of the microscope, the researcher could also determine how the hair fibers were originally obtained many years ago by examining the tip or the root of the hairs. By examining the roots of the hair, they were able to determine that both the goat and the camel hairs were shed (rather than shorn) from the animals because the roots or tips of the fibers were intact. The excavated wool fibers lacked the tips or roots because they were most likely shorn from sheep, as is usually the case.

As it is obvious that most of these fibers found in Egypt were of animal origin, we can conclude that the use of animal fibers in Egypt must have come from the Romans. The Egyptians were greatly affected by the Romans as demonstrated by their influence on Egyptian clothing and textiles.

References

1 Peacock, E.E. (1996). Biodegradation and characterization of water-degraded archeological textiles created for conservation research. *International Biodeterioration & Biodegradation* 38 (1): 49–59.

2 Faber, A. (1938). Dress and dress materials in Greece and Rome. *CIBA Review* 1: 297.

3 Batcheller, J.C. (2004). Optical and scanning electron microscopy techniques for the identification of hair fibers from Romano-Egyptian textiles. In: *Conference Proceedings Scientific Analysis of Ancient and Historic Textiles: Informing Preservation, Display, and Interpretation*, 51–56. London: Archetype.

4 Hollen, N. and Saddler, J. (1973). *Textiles*, 4e. The MacMillan Company.

5 Margariti, C., Protopapas, S., and Orphanou, V. (2011). Recent analyses of the excavated textile find from Grave 35 HTR73, Kerameikos cemetery, Athens, Greece. *Journal of Archeological Science* 38: 522–527.

6 Jake, K.A. and Mitchell, J.C. (1996). Cold plasma ashing preparation of plant phytoliths and their examination with scanning electron microscopy and energy dispersive analysis of X-rays. *Journal of Archeological Science* 23: 149–156.

7 Bergfjord, C. and Holst, B. (2010). A procedure for identifying textile bast fibers using microscopy: flax, nettle/ramie, hemp and jute. *Ultramicroscopy* 110: 1192–1197.

8 Franceschi, V.R. and Nakata, P.A. (2005). Calcium oxalate in plants: formation and function. *Annual Review of Plant Biology* 56: 41–71.

9 Barber, E.J.W. (1991). *Prehistoric Textiles.* Princeton, NJ: Princeton University Press.

10 Elsharnouby, R.M.A. Linen in ancient Egypt. *Journal of the General Union of Arab Archeologists* 15: 1–22. http://jguaa.journals.ekb.eg/article_3087_b97ae56f7bc490c84879a457e8a9ba0a.pdf (accessed 10 June 2018).

Index

Textile Fiber Microscopy: A Practical Approach, First Edition. Ivana Markova.
© 2019 John Wiley & Sons Ltd. Published 2019 by John Wiley & Sons Ltd.

www.ingramcontent.com/pod-product-compliance
Lightning Source LLC
Chambersburg PA
CBHW072225250125
20788CB00013B/136
* 9 7 8 1 1 1 9 3 2 0 0 5 0 *